Guide to
Home Security

Other All Thumbs Guides

Car Care
Compact Disc Players
Home Energy Savings
Home Plumbing
Home Wiring
Painting, Wallpapering, and Stenciling
Repairing Major Home Appliances
VCRs

ALL THUMBS

Guide to
Home Security

Robert W. Wood
Illustrations by Steve Hoeft

TAB Books
Division of McGraw-Hill, Inc.
Blue Ridge Summit, PA 17294-0850

FIRST EDITION
FIRST PRINTING

Library of Congress Cataloging-in-Publication Data

Wood, Robert W., 1933–
    All thumbs guide to home security/by Robert W. Wood;
  illustrated by Steve Hoeft.
      p. cm.
    Includes index.
    ISBN 0-8306-4166-1
    1. Dwellings—Security measures—Amateurs' manuals. I. Title.
TH9745.D85W66 1993
643'. 16—dc20   92-41243
                                                                 CIP

Acquisitions editor: Kimberly Tabor
Editorial team: Susan Wahlman, Editor
                Joanne Slike, Executive Editor
                Stacey Spurlock, Indexer
Production team: Katherine G. Brown, Director
                  Susan E. Hansford, Typesetting
                  Patsy D. Harne, Layout
                  Tara Ernst, Proofreading
                  Stephanie Myers, Computer Artist
Design team: Jaclyn J. Boone, Designer
             Brian Allison, Associate Designer
Cover design: Lori E. Schlosser
Cover illustration: Denny Bond, East Petersburg, Pa.
Cartoon caricature: Michael Malle, Pittsburgh, Pa.   ATS

The All Thumbs Guarantee

TAB Books/McGraw-Hill guarantees that you will be able to follow every step of each project in this book, from beginning to end, or you will receive your money back. If you are unable to follow the All Thumbs steps, return this book, your store receipt, and a brief explanation to:

All Thumbs
P.O. Box 581
Blue Ridge Summit, PA 17214-9998

About the Binding

This and every All Thumbs book has a special lay-flat binding. To take full advantage of this binding, open the book to any page and run your finger along the spine, pressing down as you do so; the book will stay open at the page you've selected.

The lay-flat binding is designed to withstand constant use. Unlike regular book bindings, the spine will not weaken or crack when you press down on the spine to keep the book open.

Contents

Preface *ix*

Introduction *xi*

1 Profiles of Burglars *1*

2 Outdoor Lighting & Landscaping *13*

3 Doors & Locks *32*

4 Windows *62*

5 Burglar Alarms *72*

6 Smoke Detectors & Fire Extinguishers *89*

7 First Aid *107*

Glossary *123*

Index *129*

Preface

A collection of books about do-it-yourself maintenance and repairs, the All Thumbs series was created not for the skilled jack-of-all trades, but for the average person whose budget isn't keeping pace with today's rising costs. Although major projects are still best left to professionals, as engineers and designers found easier, faster, and less expensive materials and methods for professionals, they discovered a ready market for do-it-yourselfers. Many manufacturers now design their products to meet this need. Today, because of these advances, many homeowners are upgrading their homes with little difficulty or expense.

The All Thumbs series covers such topics as home security; car care; home wiring; plumbing; painting and wallpapering; and repairing major appliances, to name a few. The easy-to-follow step-by-step instructions are accompanied by clear illustrations that put most common repairs or improvements easily within the ability of the average homeowner.

Introduction

A generation or two back it was not uncommon to find doors unlocked while the family was away. If the door was locked, the key could usually be found under the mat or hanging on a nail by the door. Front door keys were often lost, but the lock was simple enough to be opened with a bent wire. Today, that relaxed life - style is only a nostalgic memory. In about the time it takes you to read this one sentence, a burglary has been committed at a loss of about $1000. At the same time, three crimes to property were committed, and in under one minute, a fire has struck someone's home. Fires cause about 8000 deaths and cost well over two billion dollars annually. It doesn't always happen to the other guy. No home is absolutely burglarproof or fireproof. The police can't be everywhere, and the fire department is always called after the fact. But the good news is you can take a few simple and painless steps to increase the odds in your favor.

The first step is to become aware of what you're up against. Be sure to read chapter 1, "Profiles of Burglars." It introduces four typical burglars and what they look for in a target home. But more important, it tells you what situations or circumstances persuade them to pass your home altogether. Other chapters explain how to use lighting and landscaping as a deterrent, how to secure doors and windows, types of locks and their purposes, sensors and switches, burglar alarms, fire safety, and smoke detectors and fire extinguishers. The last chapter provides some important tips on first aid. Even if you only use one or two of the projects, you'll sleep better. Emergency items such as fire extinguishers, life jackets, and spare tires are seldom used, but it is comforting to know that they are there.

CHAPTER ONE

Profiles of Burglars

First of all, understand that the typical burglar is lazy and not very bright. He or she probably has committed crimes other than burglary. More than likely a burglar is armed at least with a knife, probably a gun. Burglars should be considered extremely dangerous, possibly out of control on drugs. If you surprise one in your home, and you are in the way of escape, a burglar will probably fight to get out.

These profiles are generalized, and there is always someone who doesn't fit these descriptions. Generally, most burglars are males; however, females commit burglaries as well. Daytime burglars tend to work weekdays during the hours people are at work. Most nighttime burglars prefer to rob homes on weekends. Saturday is the most popular night, followed by Friday and Sunday. About half of burglars prefer rainy, windy nights to cover up tracks and noise, while the others work only in good weather. Burglaries increase dramatically during the holidays, particularly around Christmas and New Year's, with New Year's Eve topping the list. You can see that homes are robbed day or night, any time of the year, and in any kind of weather.

The amateur

The most common burglar is an amateur. The amateur does not make a living as a burglar and might supplement the burglary income with part-time jobs or drug sales. Such a burglar is young, from early teens to mid-twenties, and probably first committed burglary between ages 11 and 15. The amateur usually avoids homes with burglar alarms and uses any hand tool available to break in, often doing a lot of damage in the process. Most of the time the house was staked out, or watched, before it was hit. Amateur burglars drive their own personal cars or stolen cars, and are often accompanied by two or three friends. Odds are that at least one of the friends will also enter the home. An amateur usually parks some distance away, then walks to the targeted home. If you see someone park on the street, between homes, and then walk down the street, apparently abandoning the car, you should call the police. They can check the license plates. If the car is suspicious or stolen, they stake out the car and wait for the driver to return. More than half of burglars hit the same place more than once. An amateur works day or night and looks for money or anything convertible to ready cash, such as guns, cameras, power tools, TVs, and stereos. The money is almost always used to buy more drugs.

The semiprofessional

The second most common type of burglar is the semiprofessional. This burglar is slightly older than the amateur, probably mid-twenties to mid-thirties. The semiprofessional probably drives his or her own car and might even park in the driveway of the target home. This burglar brings tools, such as a pry bar, screwdriver, large slip-joint pliers, and a thin-blade knife. The semiprofessional works day or night, but normally avoids homes with burglar alarm systems. The semiprofessional usually wears gloves, is very self-confident, and often talks his or her way out when found by the homeowner. When working during the day, he or she often poses as a delivery person, telephone repairer, gas-, water-, or electric-meter reader, or contractor working in the area offering cheap roofing or driveway repairs. The semiprofessional looks for money or jewelry and probably works with a partner. They might even use two-way radios for communication and police scanners to alert them if the burglary has been reported.

The neighborhood kid

The third most common burglar is the kid around ten years old, carrying a knife or screwdriver. He or she has probably shoplifted at the local convenience store, pilfered from neighbors' garages, stolen bicycles, TVs, stereos, and anything else that can be carried. Kids never wear gloves; they enter through unlocked doors or crawl through open windows to rob unoccupied homes. They operate in the daytime and probably ride bicycles.

The professional burglar

The rarest type of burglar is the professional, sometimes referred to as a cat burglar. This burglar is the most intelligent of the group. The professional is probably between 35 and 45, with the build of a jogger.

The professional tries to avoid any contact with anyone. He or she wears gloves and probably carries lock picks, a screwdriver, pliers, and a pry bar. This burglar is usually unarmed and in the house alone.

Professionals often have partners drop them off nearby and then walk to the targeted home. They enter the home sometime between the hours of 8 and 10 p.m., when the owner is out for the evening. They often communicate by two-way radio and nearly always monitor a police scanner to be alerted if the police have been dispatched. Professionals look for small, lightweight objects of value, such as money, jewelry, artwork, stamp or coin collections, etc. They are not interested in middle-income areas, but concentrate instead on very expensive homes in affluent neighborhoods. Most of these homes have alarm systems, but professionals are familiar with ways of disabling them.

To protect your home, you need to play mind games. Put yourself in the place of a burglar. What would keep you from breaking in? Your job is to convince burglars that your home is well protected, but that if they do get in, they will be caught by the police, or, more likely, left bleeding by a vicious Doberman.

One mental deterrent is to have small signs made up to stick on doors and windows. The sign might read something like "WARNING! DO NOT ENTER. AREA PATROLLED BY SILENT ATTACK DOGS" or "WARNING! DO NOT ENTER. UNCAGED RATTLESNAKE INSIDE," or "WARNING! HIGH VOLTAGE. ENTRY MAY CAUSE SEVERE BURNS OR BE FATAL." The sign doesn't have to be true. The idea is to confuse potential burglars. They're not going to be sure. They are going to ask themselves if it's worth the risk.

A similar deterrent is "AREA PROTECTED BY ACME SECURITY SYSTEMS" stickers for doors and windows—something that tells potential burglars that alarms will sound as soon as they break in. Place the stickers on all windows and doors. You can buy these signs if you don't have a real alarm system. Check the electronic stores in your area.

Another deterrent is a huge dog dish filled with a large bone or water placed near a window or door where it can be seen. The first thing a burglar will wonder is, how big is the dog that needs that big dish? Don't use a "BEWARE OF THE DOG" sign if you have a dog. Such a sign indicates you have an uncontrollable animal, and it could lead to a lawsuit if someone is hurt. Instead, use something like "AREA PATROLLED BY GUARD DOG."

Remember, burglars are lazy. They're looking for homes that can be broken into easily. Anything you can do to make it more difficult—causing confusion or increasing the noise or time it takes to break in—is added protection. A good example is a well-lit yard and gravel landscaping. Burglars hate to walk on gravel. It's very noisy.

Security check

You can begin making your house burglarproof by performing a security check:

❑ Shrubbery. Examine the shrubs and bushes near doors and windows. A burglar can hide behind them, gaining more time to break in. Trim the shrubbery to remove hiding places.

Shrubbery can provide hiding places for burglars.

❏ Doors. Check exterior doors to see if they are solid-core or metal rather than weaker hollow-core doors. Check door locks to be sure they are good-quality deadbolts.

❏ Windows. Make sure all windows have some type of antislide lock.

❏ Garage. Keep the garage door closed and locked. Keep any ladders inside the garage, inaccessible to a burglar.

❏ Yard lights. Make sure you have adequate yard lighting. Install motion-detector lights to cover areas of possible entry.

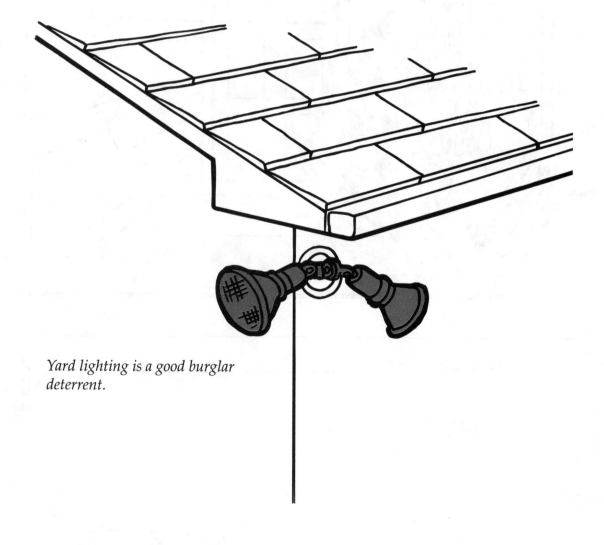

Yard lighting is a good burglar deterrent.

❑ Name signs. You have seen large decorative signs on the front of homes reading something like "THE WATSONS." Some even list the kids' names. It's a simple matter for a burglar to look up your phone number and call to see if you are home. A burglar with a car phone and a phone book could even park in front of your home and call. Never put your name on the house or mailbox.

❑ Money and jewels. Keep large amounts of cash and jewelry in a safe deposit box at a bank.

When you leave your home even for short periods of time, lock all doors and windows, unplug the phones or turn off the bell and use an answering machine, close the drapes, leave a radio on, leave a few lights on if you are returning after dark, and never leave a note on the door or in the mailbox.

When you leave your home for more than a day, *do not* tell the post office or newspaper to stop deliveries. Have a trusted friend pick up the mail, newspaper, and any advertising left on your door. Close all drapes and then go outside to make sure no one can see in; set two or three interior lights on alternating timers; leave a radio or two on moderately loud, especially near a back window; and don't forget the garbage pick-up. Have a trusted neighbor set out a little trash for you and pick up the empty can. An empty trash can left out in the afternoon when the pick-up was in the morning tells a burglar that no one is home. Have a neighbor take care of lawn mowing or snow shoveling. An unkept lawn or virgin snow on your sidewalk or driveway is a dead giveaway that no one is home. Lock all doors and windows, and turn on any alarm systems.

Another burglar deterrent is a neighborhood watch program sponsored by your local police department. Normally, such programs require a majority of the residents of the area to participate. The police do provide professional assistance and advice about crime prevention and personal property identification. They might offer free home security inspections, in which a security inspector inspects your home and provides you with a checklist of problem areas and what to do about them. The program also provides you with a personal property identification checklist where you can record items, their values, and their serial numbers.

Participate in or form a neighborhood watch program.

Often, the police department even loans an engraving tool you can use to mark any personal property that has no serial number. For valuable items such as jewelry, antiques, or artwork that cannot be marked, the police can show you how to take color photographs for identification. Always keep the list and photographs in a safe place for insurance purposes and to prove you are the owner if it is recovered by the police. You might think that you don't have much of value and that these precautions are just for the rich and famous. But soon after you start making the inventory, you'll discover that you're a lot richer than you thought.

Engrave your valuables with your social security number.

Take photographs of items that cannot be engraved.

Safety

Home safety is another part of home security. A safety checklist can make you aware of potential hazards in your home. Notice that dangers in one room might apply to other rooms as well.

Living room
- ❏ Don't overload electrical outlets or extension cords. Overloads can cause fires.
- ❏ Replace any frayed or brittle electrical cords.
- ❏ Don't run electrical cords under rugs, carpet, or in areas of traffic.
- ❏ Post emergency numbers such as fire, police, and poison control near the phone. Instruct children to call 911 in emergencies.
- ❏ Keep screen shields or heat-tempered glass doors in front of the fireplace. Have the chimney cleaned and inspected regularly.

Kitchen
- ❏ Keep pot handles turned inward and cover frying pans with screen covers or lids.
- ❏ Install good lighting over work areas.
- ❏ Mount an ABC fire extinguisher near the exit door.
- ❏ Keep household cleaners out of reach of children, and never mix bleach with oven cleaner, ammonia, or lye.
- ❏ Install ground-fault circuit interrupters (GFCI) in all receptacles within 6 feet of the sink.
- ❏ Keep pot holders, dish towels, paper towels, plastic utensils, and curtains away from the stove.
- ❏ If you smell gas, don't turn the lights on or off or use the phone. Leave the house immediately and call the fire department from a neighbor's house.

Hallways and stairs
- ❏ Have three-way light switches at the top and bottom of stairs and at each end of long hallways.
- ❏ Have smoke detectors on each floor above landings and one near the bedrooms.

❏ Make sure that stair treads have nonskid surfaces and that banisters or handrails are sturdy and stable.

Bedrooms

❏ Place a phone with emergency numbers and a lamp near the bed.

❏ Locate space heaters out of traffic and make sure they are clear of draperies, furniture, and other flammables.

❏ Don't tuck electric blankets under the mattress. Keep them smooth and flat.

❏ Remove any clutter from the floor near the bed.

❏ Place alert decals on windows of elderly or disabled persons for emergency rescue.

Children's bedrooms

❏ Make sure windows and screens are locked.

❏ Apply child-alert decals to windows to alert the fire department.

❏ Don't use space heaters in a child's bedroom.

❏ Install safety plugs in all unused electrical outlets.

Bathroom

❏ Replace regular receptacles with GFCI receptacles.

❏ Keep electrical devices such as radios, shavers, and hair dryers away from the sink or tub.

❏ Make sure shower doors are made of safety glass or heavy-duty plastic.

❏ Install a grab bar to steady anyone entering or leaving the tub or shower.

❏ Check the water temperature before entering the tub or shower.

❏ Never leave young children in the tub unattended, even for a minute.

Garage or basement

❏ Install at least one GFCI for using power tools.

❏ Install extra lighting over work areas.

❏ Make sure circuit breakers or fuses are labeled in the service panel.

❏ Never store flammable or volatile liquids in a room with a gas water heater or furnace.

❏ Keep all tools locked up or out of reach of children.

Most accidents and fires that happen in the home are preventable. We are busy enough with everyday schedules that we often get careless and overlook a safety hazard until something happens. But with a little common sense and a little effort, we can avoid being painfully reminded of simple steps we should have taken, but didn't.

CHAPTER TWO

Outdoor Lighting & Landscaping

Night burglars like to work without being seen by neighbors or anyone passing by. A few well-placed outdoor lights provide a deterrent by illuminating hiding places and points of entry. The system need not be elaborate or expensive. Lighting manufacturers have responded to the needs of the do-it-yourselfer by providing photoelectric cells that automatically turn lights on and off at dusk and dawn, motion-detector floodlights, low-voltage lighting systems, and even solar-powered lights that need no wiring at all. Solar-powered lights are more expensive than other types, but they can be installed in places where it is not convenient to run wires.

Floodlights

Automatic
light

Low-voltage
outdoor lights

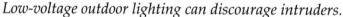

Low-voltage outdoor lighting can discourage intruders.

Shrubs and bushes can provide places for burglars to hide. The
ideal target house for burglars is one with overgrown plants, tall
hedges, or anything that provides cover while they work. Trim large
shrubs near doors and walkways. Plants with large thorns or spines,
such as roses and bougainvilleas, discourage intruders. Plant them by
windows and fences. Just make sure to keep windows clear for escape
in emergencies. Gravel areas below windows are another deterrent. It's
hard to sneak across gravel. Just pretend you are a burglar and "case"
your house some night to see how hard it would be to break in. You
have to use a little imagination—remember that a burglar on drugs
makes up in audacity what he or she might lack in intelligence.

Thorny plants discourage entry.

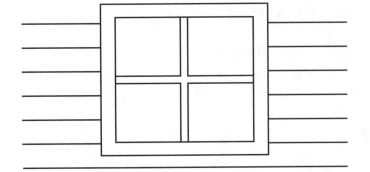

Landscape with gravel beneath windows to discourage intruders.

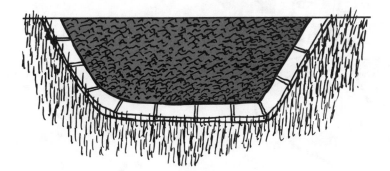

Tools & materials

❏ Screwdriver
❏ Voltage tester
❏ Wire strippers or knife
❏ Wire nuts
❏ Silicone sealant
❏ Flat spade
❏ New GFCI receptacle

Stripping wire, making connections, & using a volt meter

Steps 2-1 through 2-5 are included because some of the projects involve stripping insulation from a wire, making an electrical connection, and checking for voltage.

Step 2-1. Using a wire stripper.

Depending on the connection, you need to remove 1/2 to 3/4 inch of insulation. Match the notch in the wire stripper with the size of the wire, place the wire in the notch, and squeeze the handle. A gentle pull outward strips the insulation from the wire.

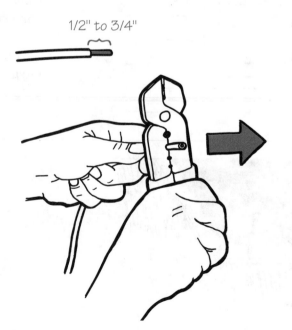

Step 2-2. Stripping wire with a knife.

If you use a knife, make the cut at an angle to avoid nicking the wire. Nicked wires are weaker and cannot carry as much current.

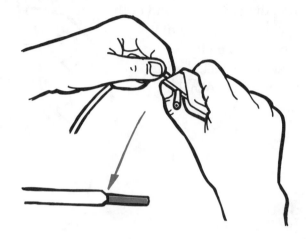

Step 2-3. Connecting wires with wire nuts.

A wire nut has threads inside and is screwed onto the bare ends of the wires. Simply hold the ends of the wires together, press the wire nut over the ends, and turn it clockwise as you would tighten a nut.

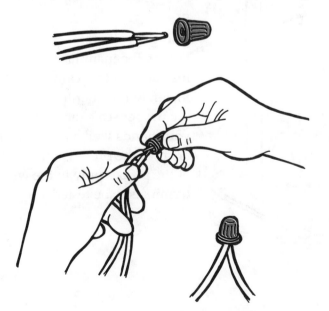

Step 2-4.
Connecting wire to a screw terminal.

First twist the bare strands of the wire to keep them from spreading. Bend the end of the wire into a clockwise loop and fit it around the screw. Tighten the screw until it makes snug contact, then tighten a half turn more to make a firm connection.

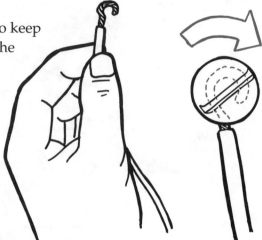

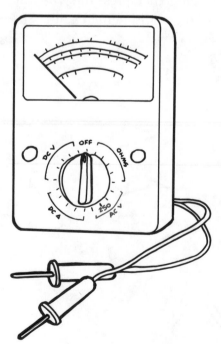

Step 2-5.
Using a volt meter.

Follow the manufacturer's instructions that came with your meter. Remember to set your meter on the proper scale (ac, 250 volts). Keep your hands well back on the insulated areas of the probes, and touch the ends of the probes only to the wires or terminals to be tested.

Installing an automatic outdoor light

An automatic light can operate by a photocell, turning on at dusk and off at dawn, or it can operate by detecting either sound or motion. Each type comes with manufacturer's instructions for adjusting and testing. Steps 2-6 through 2-11 illustrate the installation of a typical dusk-to-dawn motion detector light on an existing outlet box.

Step 2-6. Removing the old light fixture.

First turn off the switch controlling the light. For extra safety, turn off the power to that circuit at the main panel. Carefully remove any glass shade or globe; then remove the light bulb. Use a screwdriver to remove the two mounting screws holding the fixture to the outlet box. Then carefully remove the fixture from the box.

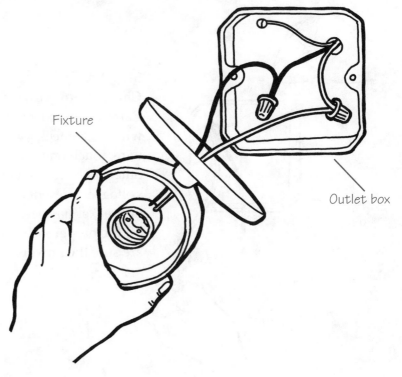

Fixture

Outlet box

Step 2-7.
Disconnecting the wires.

You will see a black wire and a white wire coming from the box, connected by wire nuts to a black wire and a white wire on the fixture. You might also find a bare ground wire to the box. Holding the insulated handles of the probes of your voltage tester, insert a probe into each wire nut to test for voltage. If there is no voltage, hold the wires and, touching only the insulation, carefully unscrew the wire nuts from each wire connection.

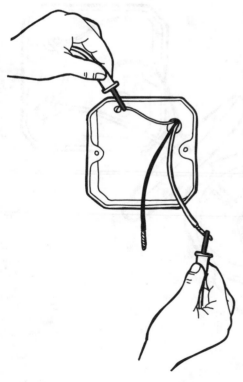

Step 2-8.
Checking for voltage.

Carefully touch a probe to the bare ends of the wires to recheck for voltage. If there is a ground wire, touch one probe to the ground wire and the other probe to each of the insulated wires. If voltage is present at either test, the wrong switch or breaker has been turned off. Go back and turn off the proper switch and repeat the test. When you are sure the power is off, go to the next step.

Step 2-9.
Connecting the wires to the automatic light.

Examine the new light. In this case you should see several wires coming from the fixture: three black wires, three white wires, and one red wire. Strip about 3/4 inch of insulation from the end of each wire. Locate the two black wires going to each lamp holder and use a wire nut to connect them to the one red wire from the motion detector. Now you have one black wire from the motion detector and three white wires, one from each lamp holder and one from the motion detector.

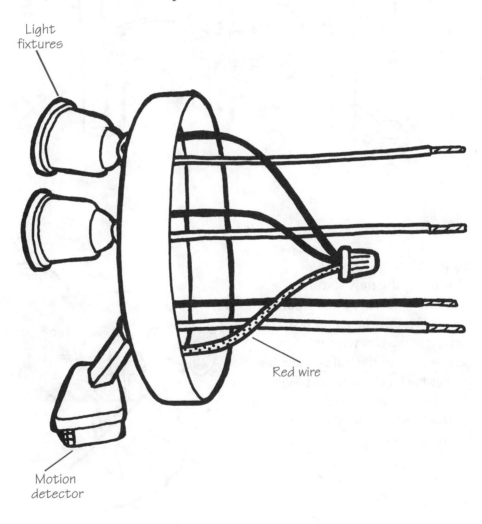

Light fixtures

Red wire

Motion detector

Step 2-10. Completing the wiring.

Feed the black and white fixture wires through the weatherproof gasket. Now use wire nuts to connect the three white fixture wires to the one white house wire. Connect the remaining black fixture wire to the black house wire.

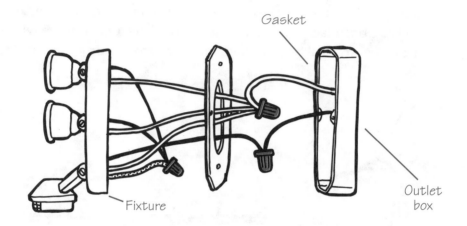

Gasket

Fixture

Outlet box

Step 2-11.
Mounting the new fixture.

Fold the wires, align the weatherproof gasket with the box, and fasten the fixture to the box with the two screws. Install the bulbs. Turn the power back on and adjust the motion detector according to the operation instructions provided by the manufacturer.

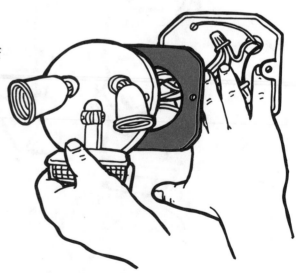

Installing low-voltage yard lights

Yard lights are excellent deterrents against intruders. They also provide safe travel on steps and walkways and enhance the beauty of the landscape. Solar-powered lights are installed by simply pushing their pointed posts into the ground. These lights have a built-in rechargeable battery that charges from the sun. The light turns on at dark and burns about seven hours from each charge. A typical low-voltage system, however, operates on 12 volts supplied by a transformer. Both systems eliminate the danger of harmful shocks from the 120-volt house system.

Low-voltage lighting kits are usually operated by a timer. The kits might have 4, 6, or 10 lights and are priced from about $30 to $60. They can be found at most hardware stores and home centers. Before you buy your light kit, study your yard. Identify a suitable location for each light and find a suitable outdoor receptacle. Outdoor receptacles must be weatherproof and protected by a ground-fault circuit interrupter (GFCI). Once you have the system planned, buy a kit that has enough lights and low-voltage cable for your needs. Steps 2-12 through 2-18 show how to install a typical low-voltage lighting kit.

Step 2-12.
Installing the transformer.
Try to locate the transformer in a protected area such as beneath a patio roof. Mount the transformer on the outside wall near the outlet. Position the transformer so that the power cord runs down, then back up to the receptacle. The power cord forms a drip loop so that rain does not run down into the receptacle. Do not plug in the transformer yet.

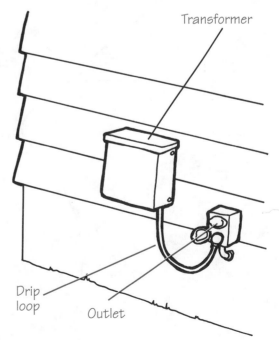

Step 2-13.
Positioning the lights and laying out the cable.
Place each light in the desired location. Then, starting at the
transformer, lay out the low-voltage cable along a route to each light.

Step 2-14.

Connecting the cable.

Open the cover of the transformer box. Now feed the cable up through the opening in the bottom of the box. The opening should be fitted with a cable clamp. Strip about 3/4 inch of the insulation from the end of both wires in the cable. Wrap the bare ends of each wire around each of the two screw terminals. Use a screwdriver to tighten the connections. Tighten the cable clamp on the bottom of the box and close the cover.

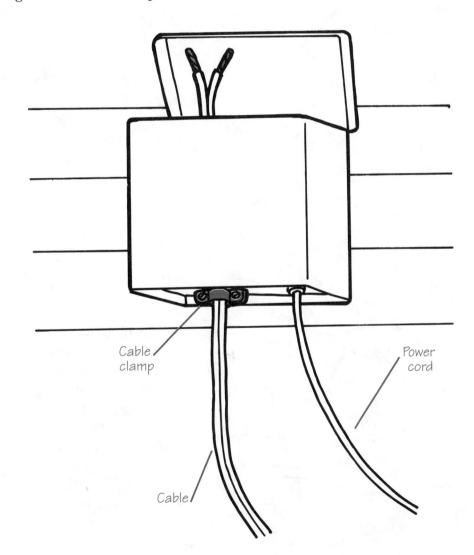

Cable
clamp

Power
cord

Cable

Step 2-15.
Connecting the lights.
Working out from the transformer, make
the electrical connections according to the
manufacturer's instructions. Some connectors have
protruding metal teeth. When the connector is pressed
in place, the teeth pierce the insulation and make the
connection to the wire.

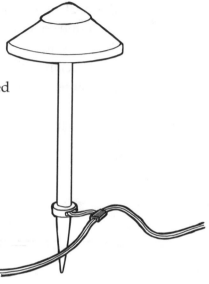

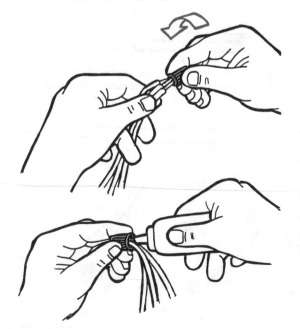

Step 2-16. Using wire nuts.
You can also use wire nuts to make the
connection. Cut the cable at the desired
location. Then strip about 1/2 inch of
insulation from the cable wires and both
light wires. Place the bare ends of two
cable wires and one light wire together
and screw on a wire nut. Tighten the wire
nut clockwise. Use silicone sealant to
weatherproof the connection.

Step 2-17. Placing the lights.

Position the lights and push the mounting stakes into the ground.

Step 2-18. Burying the cable.

To bury the cable, press a flat spade several inches into the ground with one foot. Rock the spade back and forth until a small gap is opened. Continue working until you have a narrow slit between each light. Place the cable into the slit and use your foot to press the slit closed. Now plug the transformer into the receptacle, set the timer, and check the installation.

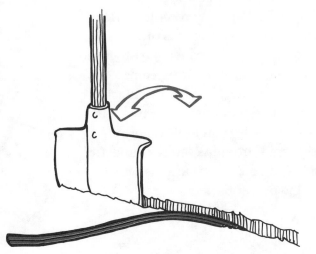

Installing a ground-fault circuit interrupter

The National Electrical Code (NEC) requires that any new receptacle installed within 6 feet of a kitchen sink, in a bathroom, or outdoors must be a ground-fault circuit interrupter (GFCI) receptacle. GFCIs should be installed in any location subject to moisture, including basements and garages. This regulation applies specifically to receptacles that you use intermittently. Receptacles serving appliances that are never unplugged, such as freezers or garage door openers, do not have to be GFCIs.

A GFCI can be installed in a regular receptacle box. It has TEST and RESET buttons that should be used about once a month to ensure that the device is working. Most GFCIs have provisions to connect to other regular receptacles, providing GFCI protection to any regular receptacle installed downstream in the circuit.

To install a GFCI receptacle, first make sure the power to the circuit is off. Remove the receptacle cover plate and unscrew the two mounting screws.

Step 2-19.
Removing the old receptacle.
Carefully pull the receptacle from the box. Recheck for voltage to make sure the power is off. You should see black, white, and ground wires just like any receptacle. The receptacle might have two sets of wires: one set from the service panel, the other set feeding the receptacles downstream. Disconnect the wires from the receptacle and bend them out of the way. If two sets of wires are present, you must determine which set comes from the service panel (the hot wire). To do so, make sure the wires are clear, and turn the power back on.

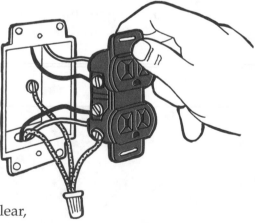

Step 2-20.
Finding the hot wire.

Use the voltage tester to determine which black wire is hot. Touch one of the probes to the ground wire and the other probe to each of the black wires. The hot wire is in the cable from the service panel. Now turn the power back off.

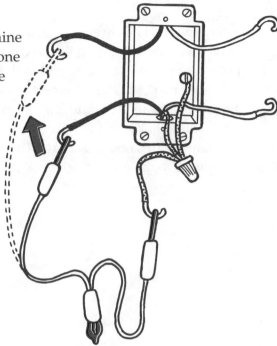

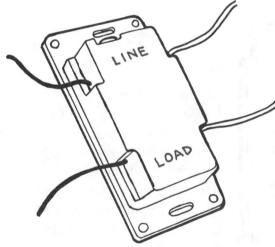

Step 2-21.
Examine the
GFCI receptacle.

The receptacle is probably prewired. Notice that two of the wires are marked LINE and two are marked LOAD.

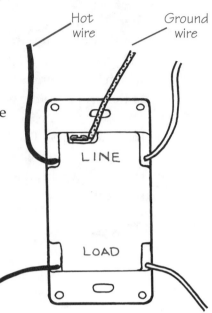

Step 2-22.

Connecting two sets of wires.

If you have two cables entering the box, connect the black hot wire from the service panel to the black receptacle wire marked LINE. Connect the white wire from the same cable to the white wire marked LINE. Connect the other black and white cable wires to the black and white wires marked LOAD. Then connect the ground wire to the ground screw.

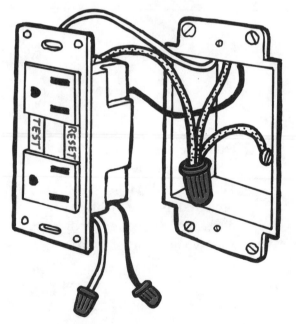

Step 2-23.

Connecting one set of wires.

If only one cable or set of wires enters the box, connect the cable wires to the receptacle wires marked LINE. Use the wire nuts to close the ends of the remaining two wires on the receptacle and fold them out of the way. Connect the ground wire to the green ground terminal on the receptacle.

Step 2-24.
Mounting the receptacle.
Carefully fold the wires into the box and install the mounting screws.
Install the cover plate and turn the power back on.

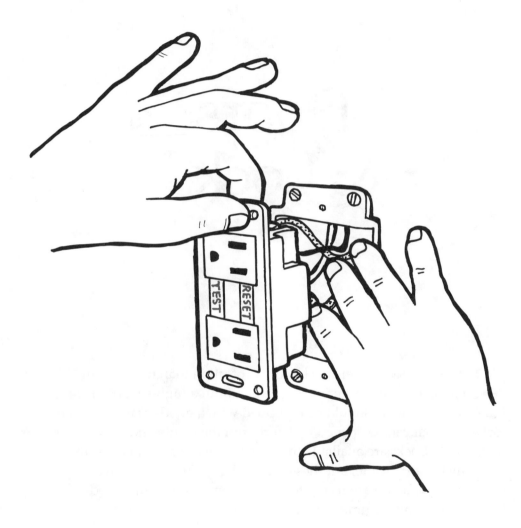

CHAPTER THREE

Doors & Locks

Solid-core doors offer much more protection than hollow-core doors, but neither are much good without a high-security lock. Entrance doors should be equipped with a good-quality deadbolt lock—one that can be opened by hand from the inside and by key from the outside. Locks are available that must be opened by a key from both outside and inside, but these can be dangerous if you have to get out in a hurry. If you are trying to escape from a fire or an armed burglar, you have no time to look for a key.

If your front door doesn't have a window, install a 180-degree, wide-angle viewer. A few homes have the door hinges installed on the outside, with the door opening outward. It is a simple matter for a burglar to remove the pins and remove the door. To solve this problem, install a pin in the door to prevent the door from being removed even if the hinges are removed.

Wide-angle viewers are easy to install.

Securing the garage door is as important as securing the front and back doors. Use good-quality steel bolts and hasps fastened to the door with bolts.

Garage doors should have good-quality hasps and locks.

Tools & materials

❏ Screwdriver
❏ Tape measure
❏ Pencil
❏ Electric drill
❏ Wood bit
❏ Large nail
❏ Hacksaw
❏ File
❏ Hole saw
❏ Utility knife

❏ Locking pliers
❏ Pliers
❏ Lipstick or grease pencil
❏ Wood chisel
❏ Hammer
❏ Masking tape
❏ Carpenter's level
❏ Toothpicks or wood dowel
❏ Wood glue

Installing a wide-angle viewer

Step 3-1. Drilling the hole.

Determine a height on the door that is convenient for members of the family. You might want to install an additional peep hole at a lower level for children or people in wheelchairs. Use a tape measure and pencil to locate and mark the center of the door at the desired height.

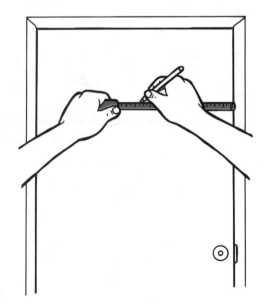

Step 3-2.
Drilling the hole.
Use an electric drill fitted with a wood bit the size of the shank of the viewer (normally 1/2 inch). Don't drill all the way through— the edge of the hole will splinter. Stop drilling when just the tip of the bit comes through the door. Now go to the other side of the door and drill back through.

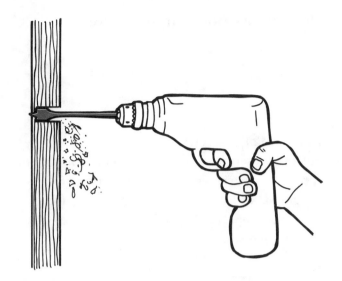

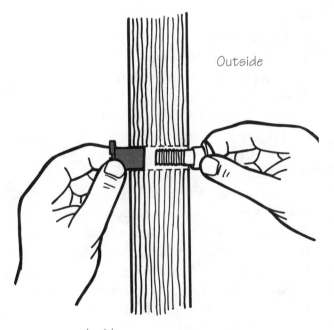

Outside

Inside

Step 3-3.
Installing the viewer.
The viewer comes in two parts— the inside part and the outside part. Determine which is which, then insert the two pieces into the hole and screw them together. Tighten the inside part by hand if it has knurled edges or use a coin if it has slots.

Pinning a door that opens out

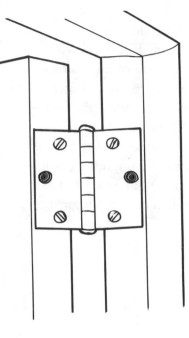

Step 3-4.
Removing the screws.
Select two screws opposite each other in the top hinge. Choose screws near the middle of the hinge. Make sure the screws line up. If not, go to Step 3-6. If they do line up, remove the two screws.

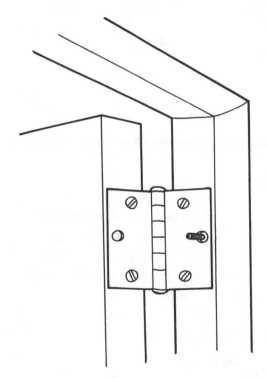

Step 3-5.
Installing the pin
when the screws line up.
Drive a large nail into the empty screw hole in the hinge attached to the door jamb. Don't drive the nail all the way in. Use a hacksaw to cut off the nail head so that about 1/2 inch of the nail protrudes from the hinge. Smooth any rough edges with a small file, then close the door. If the door doesn't want to close, tap the nail in a little further. You also might use a small bit and drill out the opposite screw hole to accommodate the end of the nail. Now do the same with the bottom hinge.

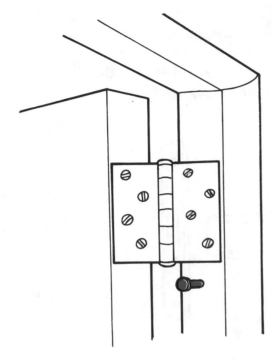

Step 3-6.
Installing the pin when the screws don't line up.
If the hinge screws don't line up, drive the nail partway into the door frame just below the top hinge. Use a hacksaw to cut off the nail head leaving about 1/2 inch protruding from the frame. Smooth any rough edges with a small file.

Step 3-7.
Completing the job.
Close the door slightly so that the nail marks the door, and drill a hole at the mark in the edge of the door. The hole should be slightly larger than the nail and just deep enough to fit the end of the nail. Now do the same just above the bottom hinge.

Installing a deadbolt lock

A good-quality deadbolt lock should have a solid, case-hardened steel cylinder guard. A rotating cylinder guard is even better. The lock also should have case-hardened screws to fasten both pieces of the lock to the door and a bolt that protrudes at least 1 inch from the edge of the door.

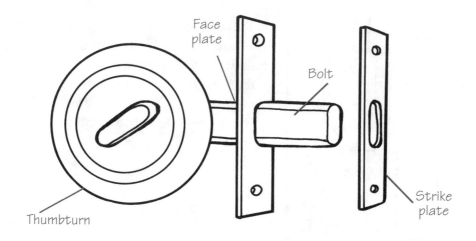

Install a good-quality deadbolt lock.

Step 3-8. Using the template.

You should find a template supplied with your new lock. The template is marked to allow for different door thicknesses, so be sure to use the marks for the size of your door. Fold the template, making sharp corners, along the lines for your door size. Position the template about 6 inches above the existing door knob and tape it to the door.

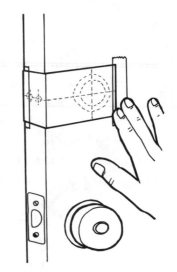

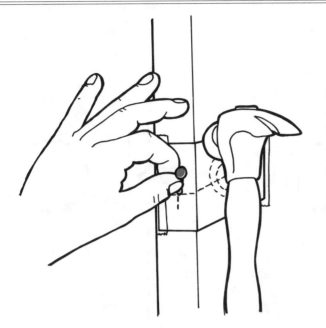

Step 3-9.
Marking the holes.

Use a hammer and a nail to mark a point through the center of the marks on the template. Mark both the bolt hole and the hole for the lock. Remove the template.

Step 3-10.
Drilling the hole for the lock.

Close the door or wedge it so that it won't move while you're drilling. Use an electric drill and a hole saw of the size recommended by the lock manufacturer. Start the drill in the nail mark for the lock. Drill slowly. Don't try to force the saw through the door. Stop once or twice to clear the wood from the saw. To keep from splintering the wood around the edge of the hole, stop drilling when the small bit breaks through the other side. Go to the other side of the door and, using the hole made by the bit, drill back through to complete the hole.

Step 3-11.
Drilling the bolt hole.
Firmly wedge the door open
to keep it from moving while
you drill. If you have a drill
guide, use it to keep the drill
bit straight and level. If not,
ask someone to guide you.
You keep the drill bit straight

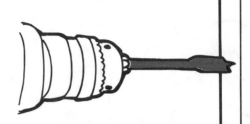

from above while your helper watches at eye level to keep it
straight horizontally. Use a wood bit of the size recommended
by the lock manufacturer. Start the tip of the bit in the nail
mark for the bolt. Apply gentle, steady pressure and drill the
hole into the hole for the lock. Blow the sawdust from
both holes.

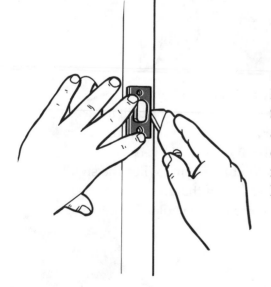

Step 3-12.
Marking for the face plate.
Slide the bolt assembly into the bolt hole
and press the face plate flush against the edge
of the door. Use a utility knife to
mark the outline of the face plate.
Remove the face plate and bolt assembly.

Step 3-13.
Using the chisel.
Use a sharp wood chisel to cut out the marked area for the face plate. Hold the chisel with the beveled side down. Be neat. Cut just to the thickness of the face plate and stay within the marks. Reinstall the bolt assembly and face plate.

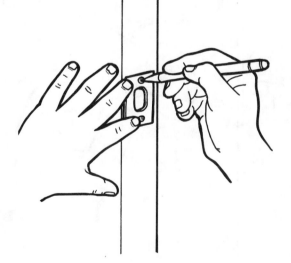

Step 3-14.
Installing the screws.
Mark the screw holes and use a small bit to make pilot holes for each screw. Now fasten the assembly in place with the screws.

**Step 3-15.
Adjusting the
lock to the door.**
Because of the various
thicknesses of doors, you
might have to shorten the
drive bar—the bar that moves
the bolt. Measure the length of the
drive bar and compare it to the thickness of
the door. The length of the drive bar should be
a little less than half the thickness of the door.
The drive bar is made up of segments. To
shorten the drive bar, use locking pliers to
grip the bar just inside the desired segment,
then use pliers to break off the segments you don't need.

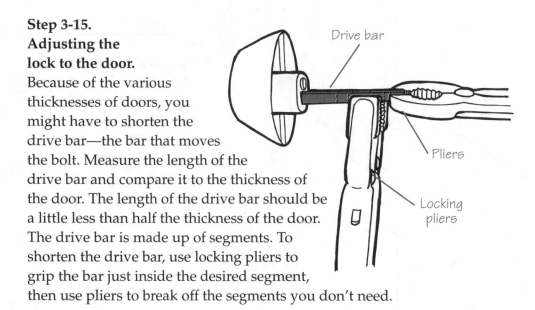

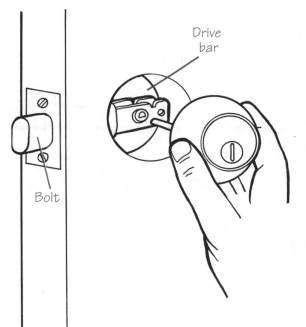

**Step 3-16.
Installing the lock assembly.**
Temporarily assemble the lock to make
sure the drive bar turns in the right
direction to operate the bolt. If it doesn't,
rotate the drive bar 180 degrees. Slide the
lock assembly into the hole from the
outside of the door. Fit the drive bar into
the drive bar hole in the bolt.

Step 3-17.
Installing the retaining plate.
From the inside of the door, install the retaining plate
and tighten the retaining screws.

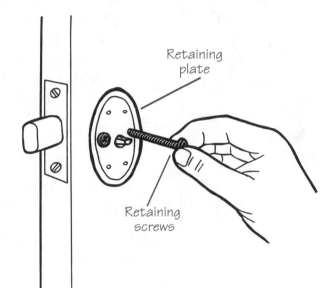

Step 3-18.
Installing the thumbturn.
Install the thumbturn by fitting
the drive bar into the thumbturn
hole and aligning the screw holes
in the thumbturn with the ones in
the retaining plate. Now insert the
mounting screws through the
thumbturn, the retaining plate, and
the bolt assembly, and screw them
into the outside part of the lock.

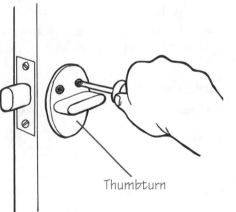

Step 3-19.
Marking for the bolt.
Use the thumbturn to extend the bolt. Now coat the end of the bolt with lipstick or a grease pencil. Retract the bolt, close the door and use the thumbturn to press the bolt against the jamb (door frame), marking the jamb.

Step 3-20.
Marking for the strike plate.
Open the door and place the strike plate against the jamb with the opening centered over the mark. Use a utility knife to mark the outline of the strike plate, cutting into the jamb to the thickness of the strike plate. Use a pencil to mark the screw holes and the opening.

Strike plate

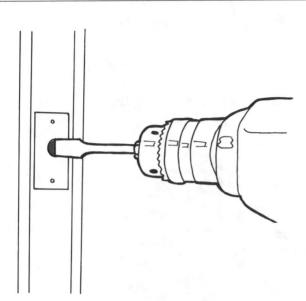

Step 3-21.
Drilling the holes
for the strike plate.
Use an electric drill with a wood
bit (the size you used to drill the
bolt hole) to drill out the marked
opening for the strike plate.
Drilling two overlapping holes
will probably do the job. Drill
only to the depth of the bolt. Use
a small bit to drill pilot holes for
the screws.

Step 3-22.
Installing the strike plate.
Use a wood chisel to cut out the outline of
the strike plate. Cut just to the thickness
of the strike plate. When the strike plate
fits flush with the jamb, fasten it to the
jamb with screws.

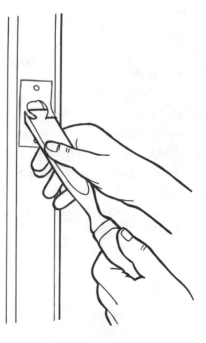

Installing a rim lock

A rim lock has a vertical deadbolt that slides into a hardened metal hasp fastened to the door jamb. It is an excellent lock because it cannot be forced or pried open. However, any lock is only as strong as the screws holding it to the door and frame.

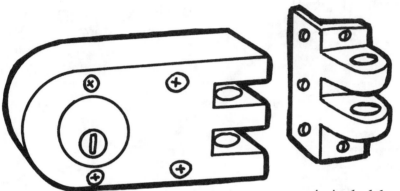

*A rim lock has a
vertical deadbolt.*

Step 3-23.
Marking the holes.
Your new rim lock comes with a template. Fold the template along the lines indicated for your size of door. Tape the template to the door with the bottom of the template about 6 inches above the existing doorknob. Use a hammer and a nail to mark the center of the cylinder hole and each screw hole. Now remove the template.

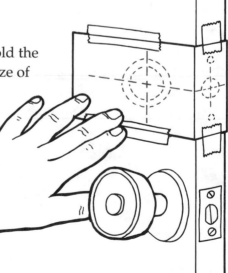

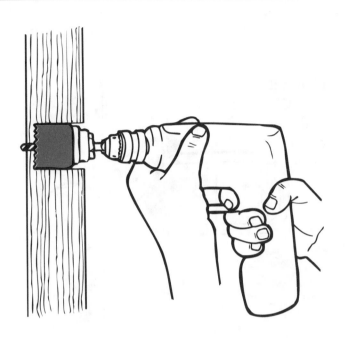

Step 3-24.
Drilling the holes.
To drill the hole for the cylinder, use an electric drill with a hole saw of the size indicated on the template. Start the bit in the hole made by the nail. Hold the drill level and use slow, steady pressure. Stop once or twice to remove the accumulated wood from the hole saw. To keep from splintering the wood, stop the drill when the bit breaks through the other side of the door. Go to the other side and drill back through to complete the hole. Use the drill and a small bit to make pilot holes for each screw. Brush or blow away any sawdust.

Step 3-25.
Installing the lock.
If necessary, adjust the length of the drive bar (see Step 3-15). Fit the cylinder assembly into the hole from the outside of the door. Use screws to fasten the retaining plate to the inside of the door. Tighten these screws. Fit the end of the drive bar into the thumbturn hole in the lock case and use screws to fasten the lock case to the door.

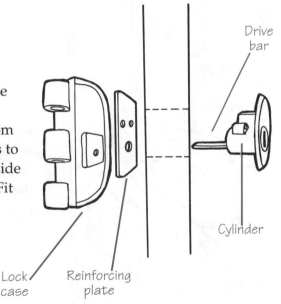

Drive bar

Cylinder

Lock case

Reinforcing plate

Step 3-26.
Marking the top
and bottom of the strike plate.
Fit the strike plate into the lock case and turn the thumbturn to hold the strike plate in place. Close the door so that the strike plate is pressed against the jamb and the door trim. Use a utility knife to mark the top and bottom edges of the strike plate on the door trim.

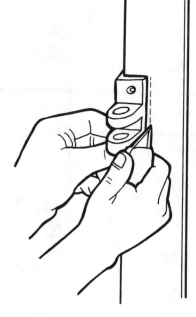

Step 3-27.
Marking the side of the strike plate.
Remove the strike plate from the case and place it between the two marks on the door trim. Now mark the door trim for the sides of the strike plate. Cut a line over on the trim to the thickness of the strike plate. You must allow for cutting in the same depth on the jamb.

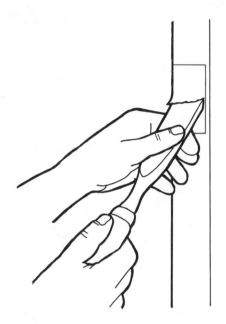

Step 3-28.
Cutting the door trim.
Use a wood chisel to cut out the wood inside the marked area. Hold the chisel with the beveled side down. Cut only to the thickness of the strike plate.

Step 3-29.
Marking the jamb.
Fit the strike plate into the cut area and use the utility knife to mark the flat part of the strike plate on the door jamb.

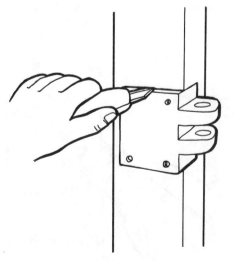

Step 3-30. Cutting the door jamb.

Use the wood chisel to cut out this area of the jamb. Fit the strike plate into the recessed area and test to make sure the lock bolt works smoothly.

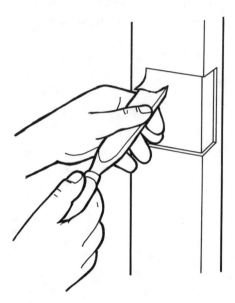

Step 3-31. Completing the job.

Mark the screw holes in the jamb and door trim and drill small pilot holes for the screws. Fasten the strike plate with screws.

Sliding glass doors

Sliding glass doors are popular with burglars because they are easy to open. You can install screws in the upper track to keep the door from being lifted out of its tracks. To keep it from sliding, you can place a broom handle in the bottom track of a closed door. You also can buy and install a security bar or a deadbolt lock.

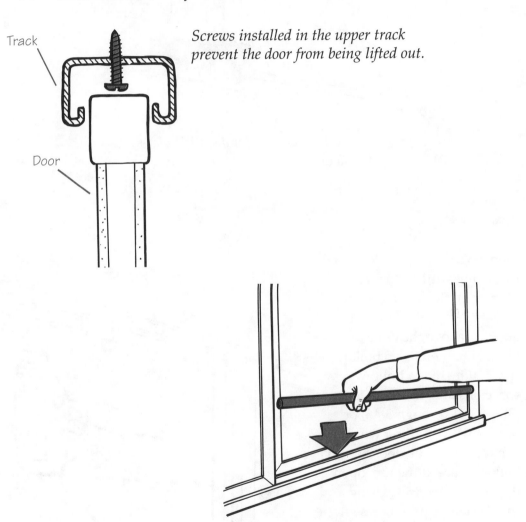

Track

Door

Screws installed in the upper track prevent the door from being lifted out.

A broom handle placed in the lower track prevents the door from being opened.

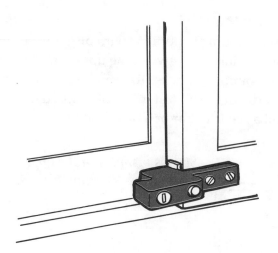

A deadbolt lock for a sliding glass door is a good safety measure.

Step 3-32.
Installing antilift screws.
First, open the sliding door and stick your finger into the opening above the door to make sure there is no hollow space. If there is a hollow space, install a deadbolt lock (Steps 3-41 through 3-46) or pins as with a sliding window (Chapter 4). If the door has no hollow space at the top, open the door all the way. Using a small bit, drill pilot holes for 1½-inch-long #12 sheet-metal screws every 10 inches or so in the center of the upper track.

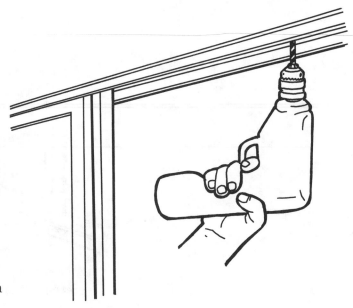

Step 3-33.
Completing the job.

Use a screwdriver to drive the screws partway into the pilot holes. Leave enough of the screw heads sticking down to keep the door from being lifted, but allow enough clearance so that the door does not rub when it is opened and closed.

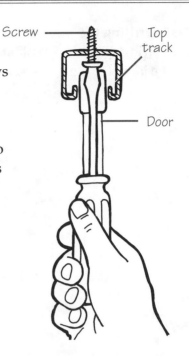

Step 3-34. Installing a security bar.

Close the sliding door, and, from the inside, hold the bar horizontal about halfway up the stationary glass panel. Place the hinge bracket against the jamb of the stationary panel. Use a pencil to mark the jamb for each screw hole in the hinge bracket.

Step 3-35. Drilling the pilot holes.
Use an electric drill and a small bit to drill pilot
holes in the jamb for the hinge screws.

Step 3-36.
Installing the hinge bracket.
Use a screwdriver to fasten the hinge
bracket and bar to the jamb.

Step 3-37. Adjusting the bar.
Place the saddle bracket on the
end of the bar, and, holding
the bar level, adjust the bar
so that the saddle bracket
is pressed against the edge of
the sliding door.

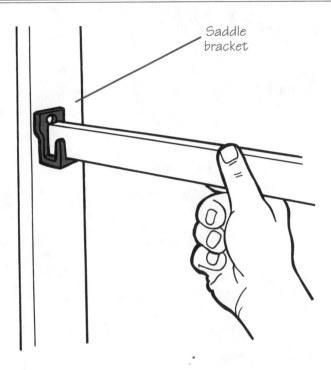

Saddle
bracket

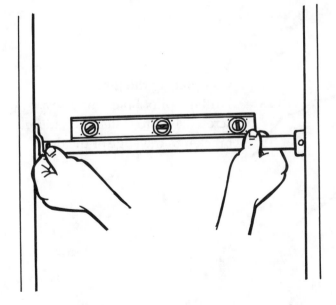

Step 3-38.
Leveling the bar.
Hold the saddle bracket
against the edge of the door
with one hand and use a
carpenter's level with the other
to make sure the bar is level.

Step 3-39.
Marking the holes.

Use a pencil to mark the edge of the door for the screw holes in the saddle bracket.

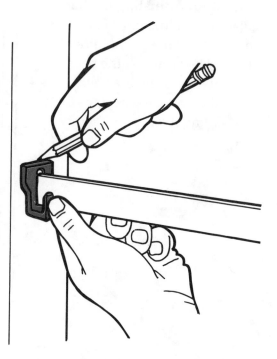

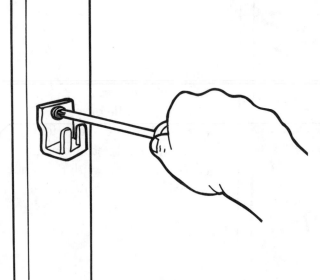

Step 3-40.
Completing the job.

Drill the pilot holes and use a screwdriver to fasten the saddle bracket to the edge of the door.

A deadbolt lock is mounted on the sliding door. The deadbolt slides into one or two holes in the stationary panel. This lock keeps the door from sliding or being lifted out of its tracks.

Step 3-41.
Marking the pilot holes.
You're going to fasten the lock to the edge of the sliding door. Close the sliding door and place the lock against the edge of the door just above the lower track. Use a nail and hammer to gently tap a starting point for the bit at the center of each screw hole in the lock.

Step 3-42.
Drilling the pilot holes.
Use an electric drill and a small bit to drill pilot holes for each screw.

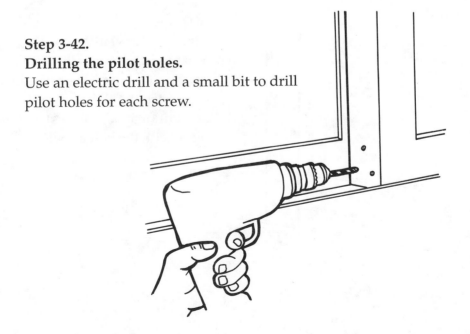

Step 3-43.
Marking the bolt hole.

Coat the end of the bolt with lipstick or a grease pencil and temporarily fasten the lock to the door. Make sure the door is closed all the way, then slide the bolt open and closed a couple of times to press the end against the stationary panel. The lipstick or grease pencil will mark the place for the bolt hole.

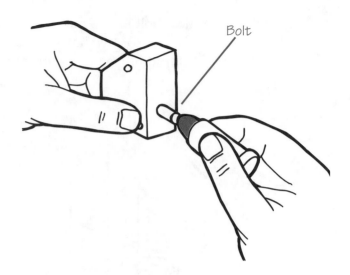

Bolt

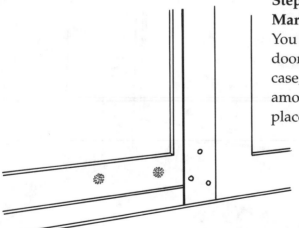

Step 3-44.
Marking for ventilation.

You might occasionally want to lock the door partly open for ventilation. In this case, open the sliding door the desired amount (maybe 4 inches) and mark this place with the bolt as before.

Step 3-45.
Marking a
starting point for the drill.
Remove the lock from the edge of the door. Use the hammer and nail to gently tap a starting point for the bit at the center of the lipstick marks for each bolt hole.

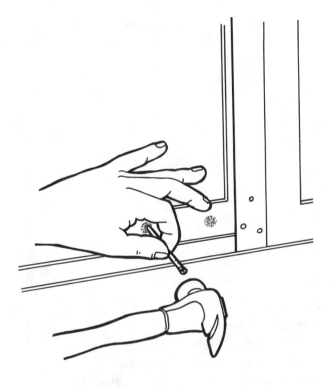

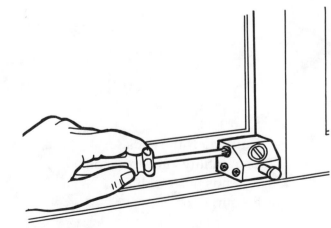

Step 3-46.
Completing the job.
Use an electric drill and a bit a little larger than the size of the bolt to drill each bolt hole. Fasten the lock to the edge of the door and test the lock.

If you make a mistake in drilling screw holes in wood, you can
repair the holes by following Steps 3-47 through 3-50.

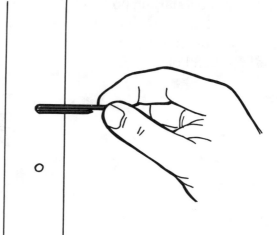

Step 3-47.
Using toothpicks.
Dip the ends of two or three toothpicks
in wood glue and gently tap them, one
at a time, into the hole. Let the glue dry.

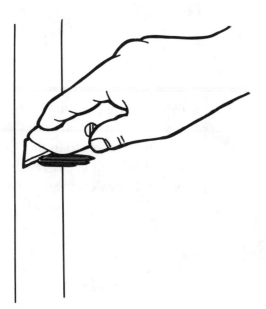

Step 3-48.
Cutting off the ends.
Use a utility knife to cut off
the ends flush with
the surface.

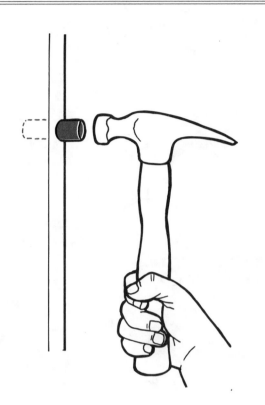

Step 3-49.
Using a wood
dowel for larger holes.
Dip the end of a wood dowel in
wood glue and tap it into the hole.
Let the glue dry.

Step 3-50. Using a chisel.
Use a sharp wood chisel to cut off the
end flush with the surface.

CHAPTER FOUR

Windows

Most burglars gain entry to homes through unlocked windows. Keep your windows locked. Several types of antislide locks are available for sliding windows. Broom handles or metal pins also can be used. Screws or wood dowels can be installed in the upper track to prevent the windows from being lifted from their tracks. Double-hung windows can be secured by bolt-type key locks or pins. Keep in mind you might have to use the window in an emergency.

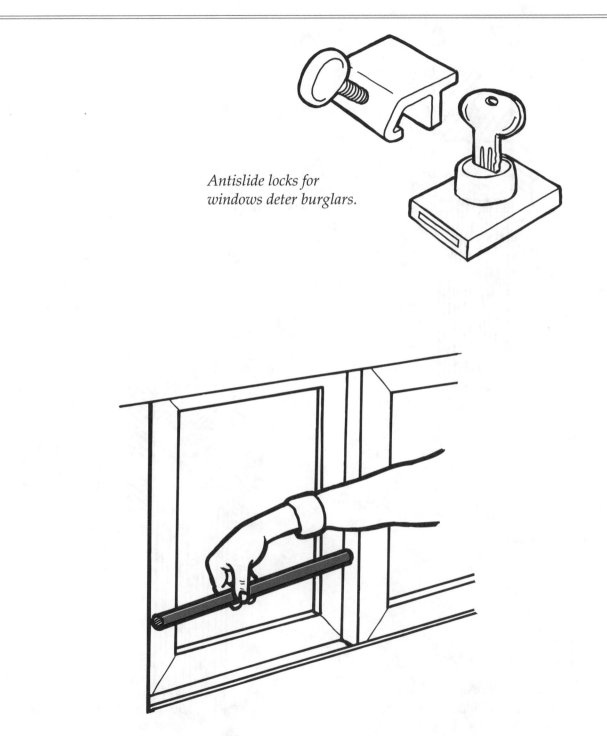

Antislide locks for windows deter burglars.

A broom handle keeps the window from opening.

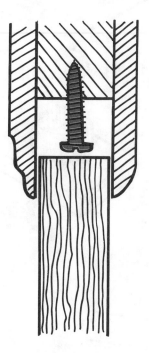

Screws installed in the upper track keep the window from being lifted out.

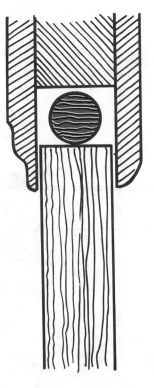

A wood dowel installed in the upper track also prevents the window from being lifted out.

Tools & materials

❏ Electric drill
❏ Small bit
❏ Sheet- metal screws
❏ Screwdriver
❏ Hammer and nail
❏ Nail or eye bolt (pin)
❏ Lipstick or grease pencil

Securing a sliding window

Steps 4-1 through 4-3 show you how to install antilift screws or dowels.

Step 4-1. Drilling the pilot holes.

Use an electric drill and a small bit to drill pilot holes in the upper track above the sliding window.

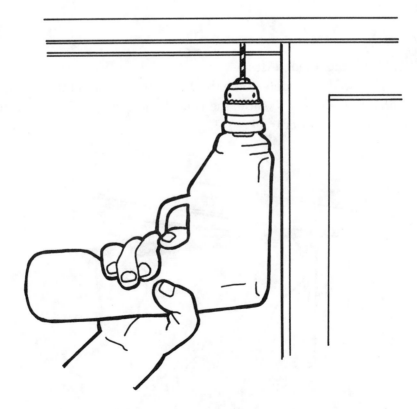

Step 4-2. Installing the screws.

Use a screwdriver to screw sheet- metal screws into the holes. Leave the screw heads sticking down just far enough to clear the window.

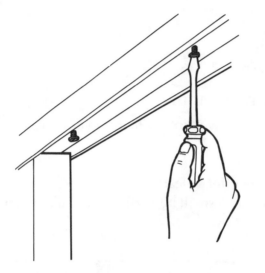

Step 4-3. Installing a wood dowel.

To install a wood dowel, determine the appropriate diameter that would allow the window to slide, then cut the length to fit the window and slide it into the upper track.

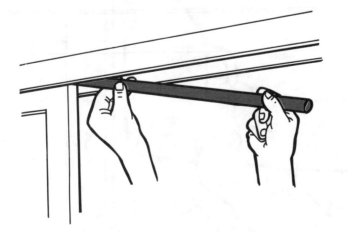

Steps 4-4 and 4-5 show you how to install pins.

Step 4-4.
Marking a starting point for the drill.
Close the window all the way. Use a hammer and a nail to tap a starting point for a small hole in the upper track above each end of the sliding window. The size of the hole and the pin depend on how much space you have to drill in the track. For the pin, select a nail or eyebolt of the proper size.

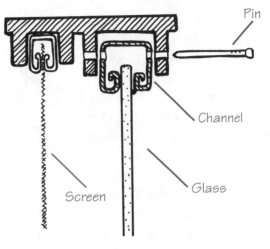

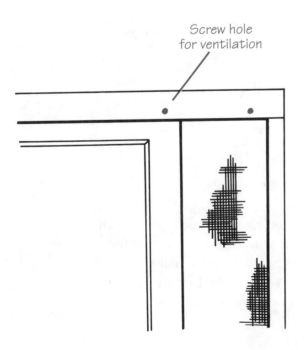

Step 4-5.
Drilling the holes.
Use an electric drill and a bit slightly larger than the pin to drill the holes. The hole should go through both sides of the track and through the sliding part of the window. Be careful not to break the glass or punch on through to the screen. For ventilation, open the window no more than 4 inches and drill another hole.

Securing double-hung windows

Step 4-6. Installing pins.
Close the window. Select a large nail
or eye bolt to be used as a pin. Use an
electric drill and a bit slightly larger
than the nail or eyebolt. Now drill a
hole through the top frame of the
lower window and partway into the
frame of the upper window. Drill
a similar hole at the other side of
the window for the other pin.
Angle the holes downward
slightly to keep the pins from
falling out. Install the pins to
secure the window. For ventilation, open
the windows no more than 4 inches and
drill another set of holes.

Steps 4-7 through 4-12 show you how to install a bolt lock on double-
hung windows.

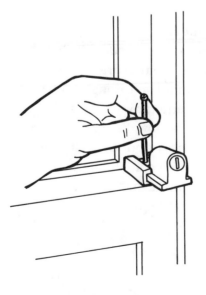

Step 4-7.
Marking the screw holes.
Close the window and place the lock on the
top edge of the lower window. Be sure the lock
is clear of the window frame so that the
window slides easily. Use a nail to mark the
center of the screw holes in the edge of the
lower window. Remove the lock.

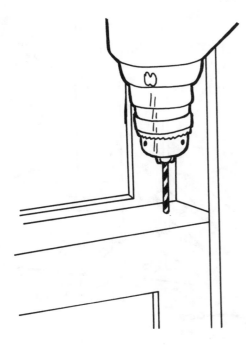
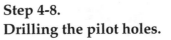

Step 4-8.
Drilling the pilot holes.
Use an electric drill and a small bit to drill pilot holes for the screws.

Step 4-9.
Marking for the bolt hole.
Coat the end of the bolt with lipstick or a grease pencil and temporarily fasten the lock to the edge of the window. Slide the bolt a couple of times to mark the upper window. To provide for ventilation, open the window no more than 4 inches and operate the bolt to make another mark. Remove the lock.

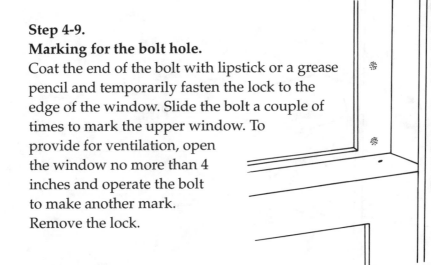

Step 4-10.
Marking for the strike plate.

Place a strike plate on the mark on the upper window. Center the bolt hole in the strike plate over the center of the lipstick mark. Use a pencil to mark the bolt hole and each screw hole. Repeat this step for the other strike plate used for the ventilating position.

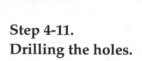

Step 4-11.
Drilling the holes.

Using a bit slightly larger than the bolt, drill a bolt hole for both strike plates. Use a small bit to drill pilot holes for each screw hole.

Step 4-12.
Completing the job.
Screw both strike plates in place and remount the lock.

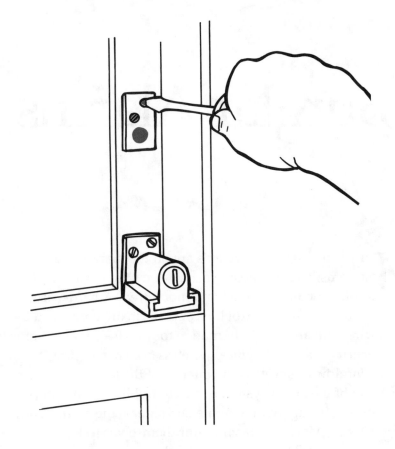

CHAPTER FIVE

Burglar Alarms

Many types of alarm systems are on the market. The best systems, and the most expensive, are the hard-wired ones professionally installed and monitored. A hard-wired system is one where the sensors, control box, and alarm are connected by electrical wires. An attractive alternative for the do-it-yourselfer is the wireless security system. A typical wireless system can protect the doors and windows, as well as the interior of the home.

The brain of the alarm system is the control box. It monitors the various sensors throughout the home and signals the alarm if any sensor is tripped. Most systems monitor themselves to let you know they are operational. The basic parts of a door or window alarm are a magnet, a switch, and a transmitter. The magnet is mounted to the door or window in close proximity of the switch. When in place, the magnet keeps a spring-loaded switch closed, allowing current to flow. When the magnet is moved, the switch opens, sending a signal to the control box that triggers the alarm.

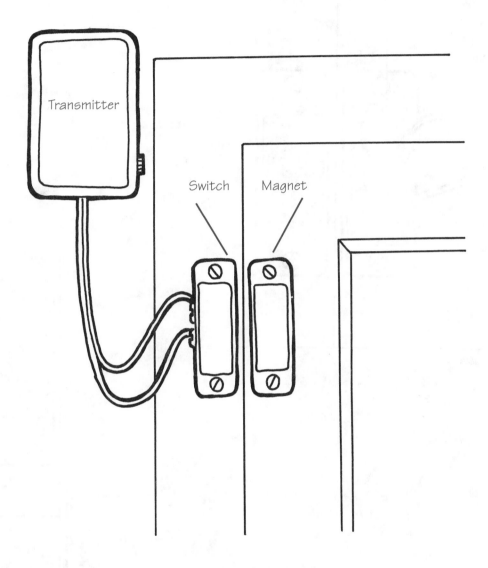

The basic parts of an alarm.

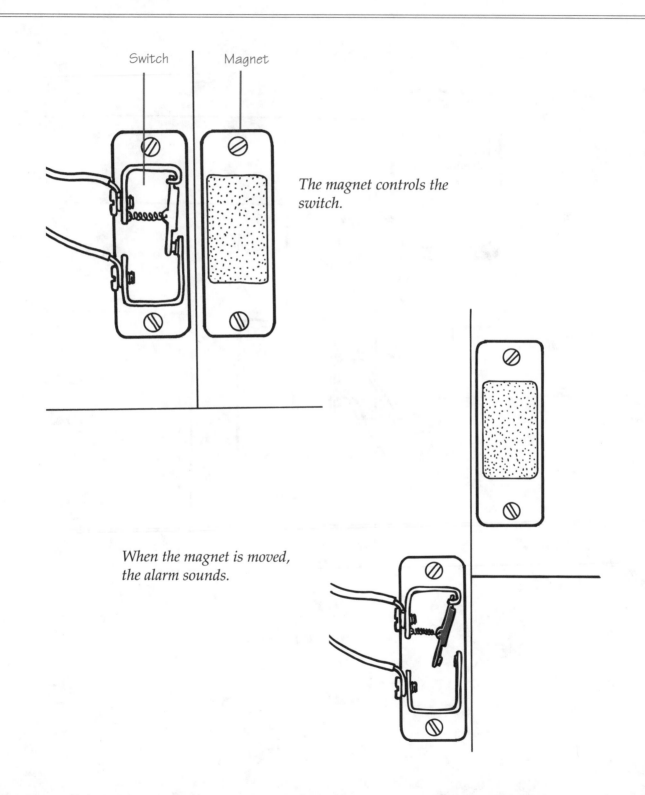

Switch   Magnet

The magnet controls the switch.

When the magnet is moved, the alarm sounds.

Another choice for the homeowner is to install self-contained alarms on doors, windows, and motion alarms for space protection. Motion-sensing alarms can be fastened to a wall with a couple of screws, or they might look like small, inconspicuous table radios that can be placed anywhere. The alarm can be battery-operated or plugged into a nearby receptacle. Battery units sound an alarm while plug-in units can turn on a lamp and sound an alarm if someone passes within range.

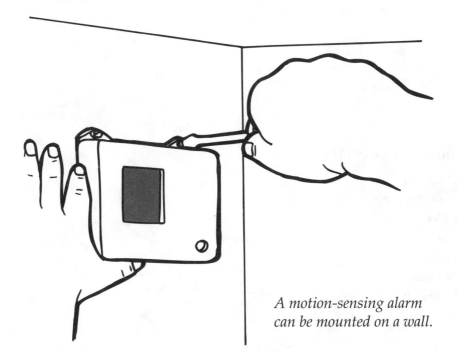

*A motion-sensing alarm
can be mounted on a wall.*

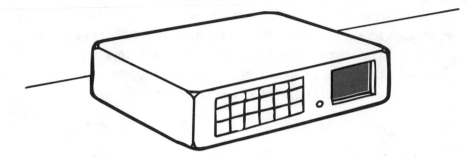

Some motion-sensing alarms look like small radios.

Tools & materials

- ❏ Screwdriver
- ❏ Hammer and nail
- ❏ Electric drill
- ❏ Small bit
- ❏ Masking tape
- ❏ Pencil
- ❏ Carpenter's level
- ❏ Toothpick or paper clip

Installing a self-contained alarm

Step 5-1. Hanging a doorknob alarm.
You can buy a doorknob alarm at most home centers. With a new battery installed, simply hang the alarm on the doorknob and turn it on. The alarm goes off if anyone touches the door. This alarm is an excellent idea for travelers.

Installing a door or window alarm

Step 5-2. Using the template.

Some alarms are mounted with screws or double-sided tape. To mount the alarm to a door using screws, use the template that comes with the alarm. Tape the template to the door so that the edge of the template is aligned with the edge of the door.

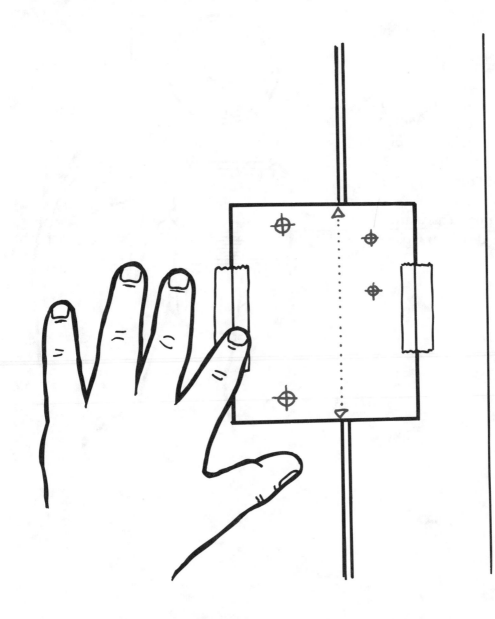

Step 5-3.
Marking starting points for the drill bit.
Use a hammer and a small nail to tap a
starting hole in the center of the outline for
each screw. Then remove the template.

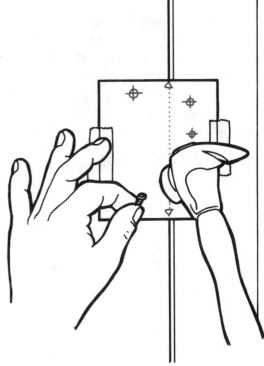

Step 5-4.
Drilling the pilot holes.
Use an electric drill and a small bit to drill pilot holes
for the screws. Fasten the alarm to the door with
screws and install a fresh battery.

Step 5-5.
Installing the magnet.
Place the magnet on the door trim so that it is in the proper position with the alarm. Mark the center of the screw holes and drill the pilot holes. Fasten the magnet to the door trim and test the alarm.

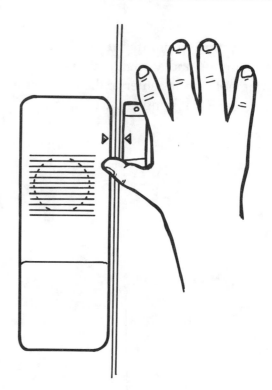

Installing a wireless alarm system

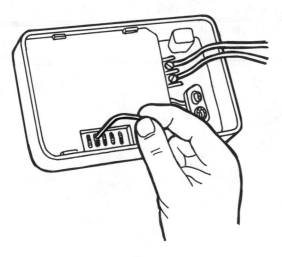

Step 5-6.
Programming the control box.
Open the cover on the control box. Use a toothpick or a straightened paper clip to set the code numbers on the dipswitch (a series of tiny switches) according to the manufacturer's instructions.

Step 5-7. Leveling the control box.
At the desired location (near a receptacle), use a level to draw a faint pencil mark for the top of the template.

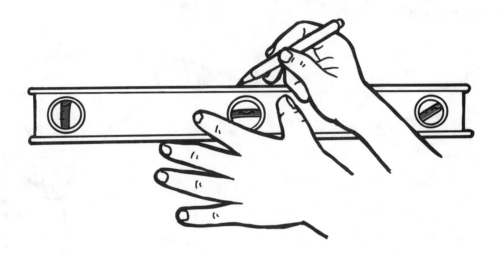

Step 5-8. Marking the template.
Place the template on the mark and use a small nail to mark the center of the outline of the screw holes. Remove the template.

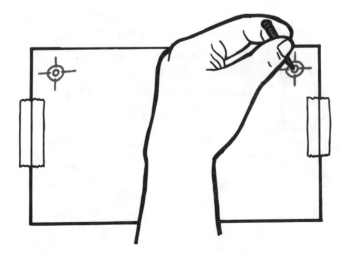

Step 5-9. Drilling pilot holes for the anchors.
Use an electric drill and a small bit to drill pilot holes for the screw anchors.

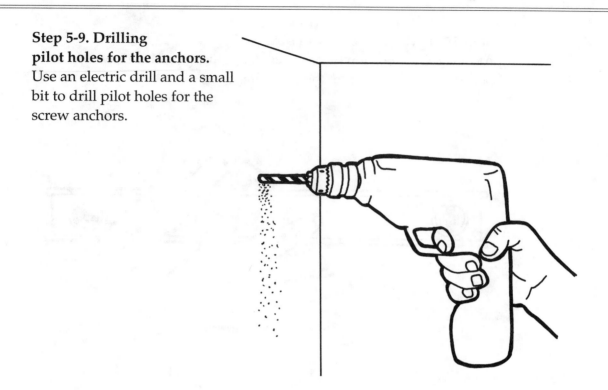

Step 5-10. Installing the anchors and screws.
Use a hammer to gently tap the anchors into the holes so that they are flush with the surface. Install the screws. Leave the screw heads extended about 1/8 inch— enough to fit into the slotted holes in the back of the control box.

Anchor

Screw

Step 5-11.
Mounting the control box.
Fit the holes in the back of the control box over the mounting screws and pull down on the box to slide the screws into the slots in the holes.

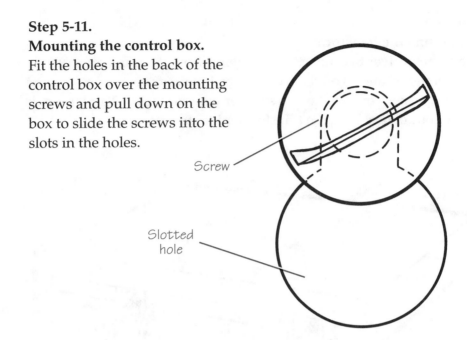

Screw

Slotted hole

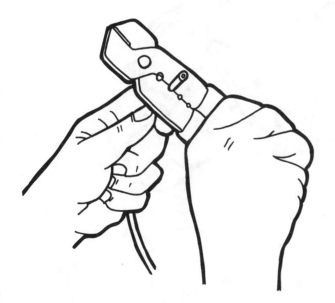

Step 5-12.
Stripping the wires.
Use wire strippers to strip about 1/2 inch of insulation from the end of each wire from the transformer. Twist the bare ends to keep the strands together.

Step 5-13.

Connecting the transformer.

Use a screwdriver to connect the two wires from the transformer to the transformer terminals on the control box. If your system has a separate indoor speaker, connect the two speaker wires to the speaker terminals on the control box. Plug the transformer into the receptacle.

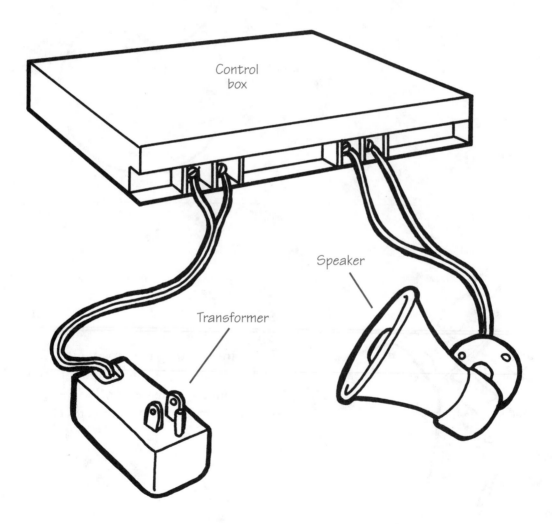

First Aid Kit Shopping List

- ❏ Adhesive tape
- ❏ Gauze roller bandage (roll about 15 feet long)
- ❏ Tweezers
- ❏ Calamine lotion
- ❏ Antibiotic spray
- ❏ Reusable ice pack
- ❏ Sterile gauze bandage (4-inch pads)
- ❏ Adhesive bandages (various sizes)
- ❏ Cotton swabs
- ❏ Small scissors
- ❏ Antibiotic ointment
- ❏ Rubbing alcohol

Fire Safety Checklist

- ❏ Plan two escape routes from each room in your home. Designate a place outside, away from the house, for everyone to meet.

- ❏ Hold fire drills and count heads at the designated place of assembly. Make everyone show up.

- ❏ Make sure that anyone using a window for escape can unlock the window, remove the screen, and safely reach the ground.

- ❏ Store chain ladders next to the upstairs escape windows.

- ❏ If you are in an apartment or hotel, notice the locations of the fire exits. Count the number of doors between your apartment and the exits.

- ❏ If a fire breaks out, don't use the elevator. A power failure could trap you inside.

- ❏ When traveling, purchase a portable smoke detector to hang on your door.

Home Security Dos and Don'ts

O Do expect a burglar to be armed and extremely dangerous.

O Do demand identification from anyone, such as a service technician or police officer, who comes to your door requesting entrance into your home. Call the company or the local authorities to verify employment and purpose of the visit.

O Do perform frequent security checks. Examine shrubs near doors and windows. Can a burglar hide behind them?

O Do check exterior doors. Are they solid-core or metal, or are they weaker hollowcore? Are door locks good-quality deadbolts? Do the windows have some type of antislide locks? Do sliding glass doors have security dead bars or deadbolt locks? Are ladders inside and inaccessible to burglars? Is the garage door locked? Is the yard adequately lit?

O Do install motion-detector, dusk-to-dawn, and low-voltage lights along the sidewalk and entranceway.

O Do place a large dog dish or big chew bone by the door or within sight of a window.

O Do consider landscaping with gravel around the house, particularly under windows, to make an approach noisy. Plant bushes with large thorns and spikes to discourage intruders.

O Do post house numbers that are at least 3 inches high where they are clearly visible for identification by emergency vehicles.

O Do participate in or start a neighborhood watch program.

O Do contact the police and ask if they offer free home security inspections.

O Do engrave personal property with your social security number or as recommended by the local police.

O Do practice emergency escape drills with the entire family.

O Do keep windows clear for escape in emergencies. Keep an emergency escape ladder handy or install a fire escape as recommended by your local fire department.

O Don't post a "Beware of Dog" sign. It indicates an uncontrollable animal and could lead to a lawsuit. Instead, post "Area patrolled by guard dog."

O Don't install doors with hinges on the outside. The pins can be removed easily, allowing removal of the door.

O Don't have the mail or newspaper delivery stopped while you're away. Ask a trusted friend or neighbor to pick it up for you. Also, have the lawn mowed or the snow shoveled.

O Don't post your name on the house or mailbox. A burglar can easily get your phone number and call to see if you are home.

From All Thumbs Guide to *Home Security* by Robert W. Wood.
© 1993 by TAB Books, a division of McGraw-Hill, Inc.

Step 5-14.
Programming the transmitters.
Remove the back cover of the transmitter and use a toothpick or straightened paper clip to enter the numbered code according to the manufacturer's instructions. Install a new battery.

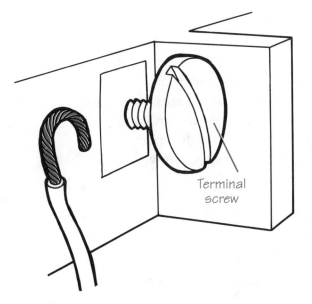

Step 5-15.
Connecting the sensor.
Strip about 1/2 inch of insulation from each wire, twist the bare ends to keep the strands together, and wrap each end clockwise around the terminal screws on the sensor. Tighten the screws with a screwdriver. Strip the other ends of each wire and connect them to the terminal screws on the transmitter.

Terminal
screw

Mounting the transmitter

Step 5-16.
Marking the screw holes.
Place the back cover on the wall near the door trim. Align the cover up and down with the door trim. Use a pencil to mark the outline of each screw hole. Remove the cover.

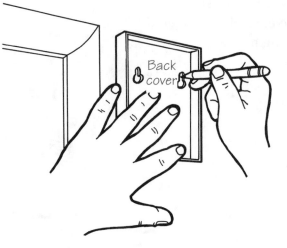

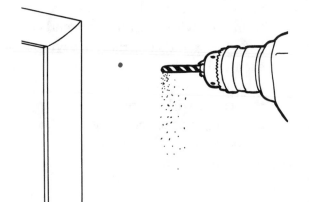

Step 5-17.
Drilling the pilot holes for the anchors.
Use an electric drill and a small bit to drill pilot holes for the screw anchors.

Step 5-18.
Installing the anchors.
Use a hammer to gently tap
the anchors into the holes.

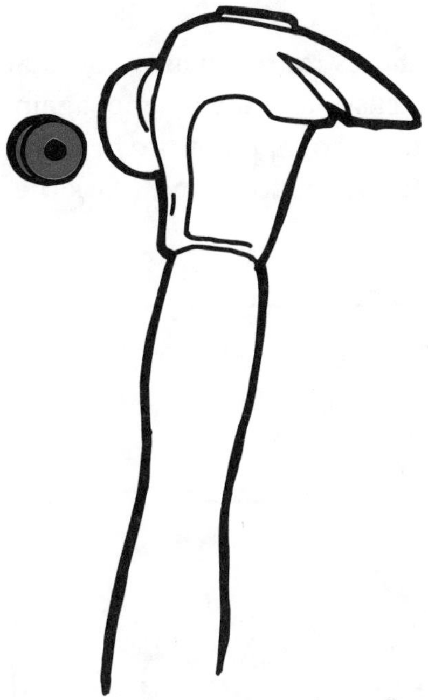

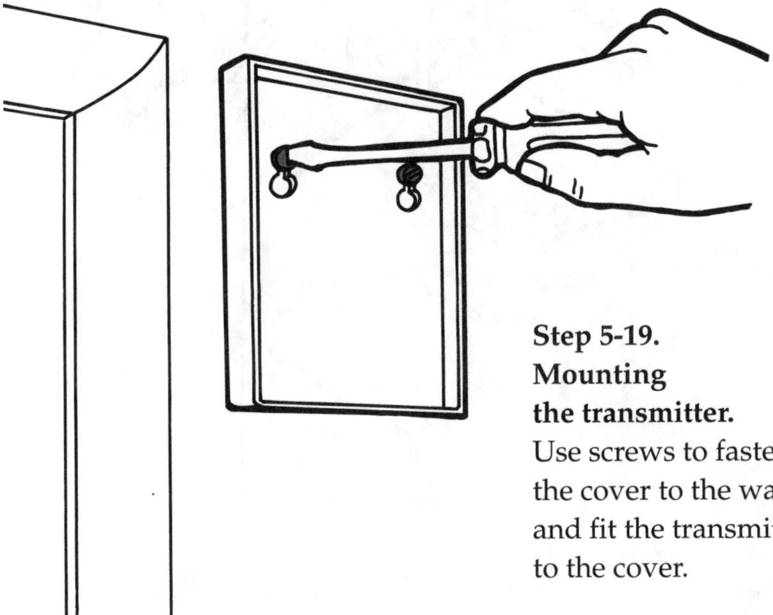

Step 5-19.
Mounting
the transmitter.
Use screws to fasten
the cover to the wall
and fit the transmitter
to the cover.

Step 5-20.
Marking holes for the sensor and magnet.
Hold the sensor and the magnet against the door trim and the door. Position them so that they are in close proximity, according to the manufacturer's instructions. Mark the screw holes with a pencil.

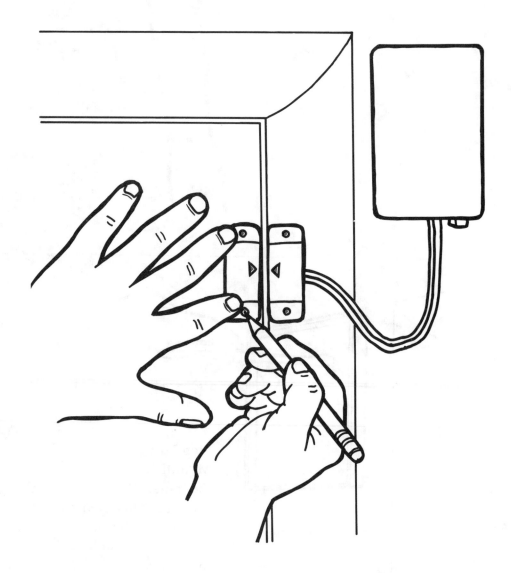

Step 5-21.
Mounting the sensor and magnet.

Use an electric drill and a small bit to drill pilot holes for the mounting screws. Fasten the sensor to the door trim and the magnet to the door. Now turn on the control box and test the system. The same procedure applies to garage doors and double-hung and sliding windows.

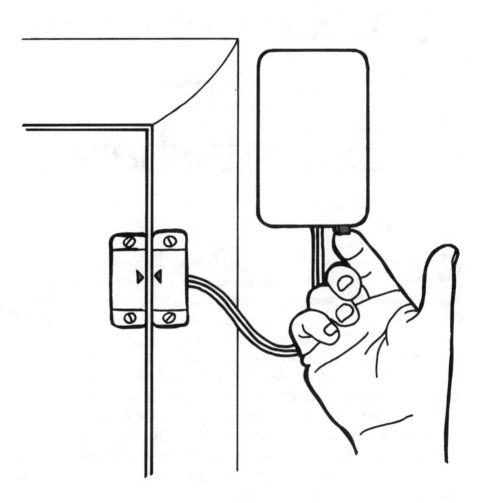

CHAPTER SIX

Smoke Detectors & Fire Extinguishers

S ome city ordinances now require smoke detectors to be installed in new construction. As a general rule, smoke detectors should be mounted on the ceiling or on the wall 6 to 12 inches from the ceiling. They should not be mounted in corners or in the path of ventilation. Install one just outside each bedroom, and, if the house has more than one level, install one on each level and above the stairway. These home fire alarm devices are available in battery-operated or direct-wired models. The direct-wired models are permanently wired into one of the house circuits. The battery-operated detectors operate independently of the home electrical system. These units normally are powered by a single 9-volt alkaline battery that should be replaced once a year. Most models have a test button and beep periodically to signal that the battery is getting weak.

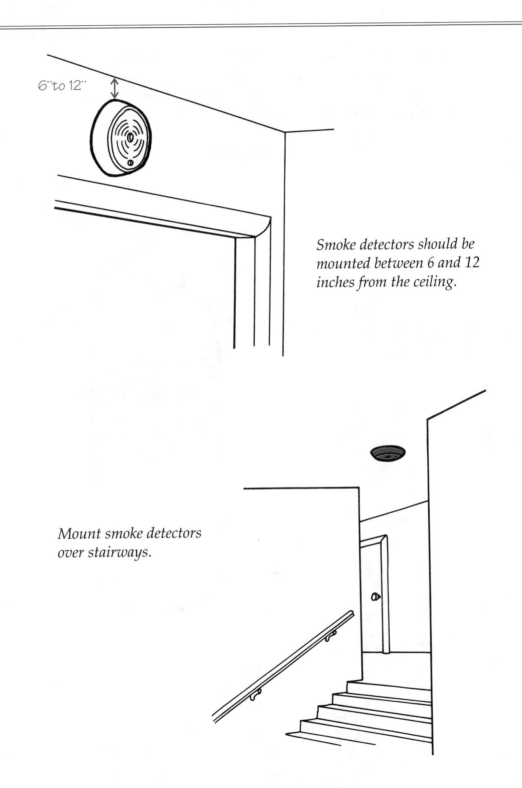

6" to 12"

Smoke detectors should be mounted between 6 and 12 inches from the ceiling.

Mount smoke detectors over stairways.

The preferred fire extinguisher for homes is the ABC-class extinguisher. It is effective against wood and paper, gasoline, and electrical fires. A typical home fire extinguisher should hold about 7 pounds of pressurized dry chemical and last about 20 seconds. However, to be of any use, it must be fully charged and accessible, and you must know how to use it without referring to the instructions.

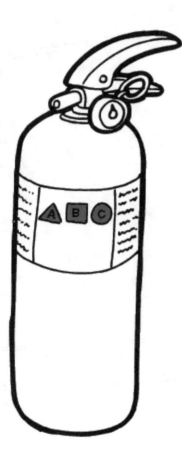

Use an ABC-class fire extinguisher in your home.

Fire safety checklist

❑ Draw a floor plan of your home. Show the windows, doors, and any stairways. Plan two escape routes from each room. Go over the routes with your family and choose someone to help the children and elderly or disabled. Be sure to designate a place outside, away from the house, for everyone to meet.

❑ Hold fire drills and count heads at the designated place of assembly. Make everyone show up.

❏ Make sure that anyone using a window for escape can unlock the window, remove the screen, and safely reach the ground.

❏ If the windows are on a second or third floor, buy chain ladders and store them next to the upstairs escape windows. These escape ladders are available at fire-equipment sales companies listed in the yellow pages of your phone book The ladders come in two sizes— 15 feet and 20 feet—and cost about $50.

❏ If you are in an apartment or hotel, notice the locations of the fire exits. Count the number of doors between your apartment and the exits.

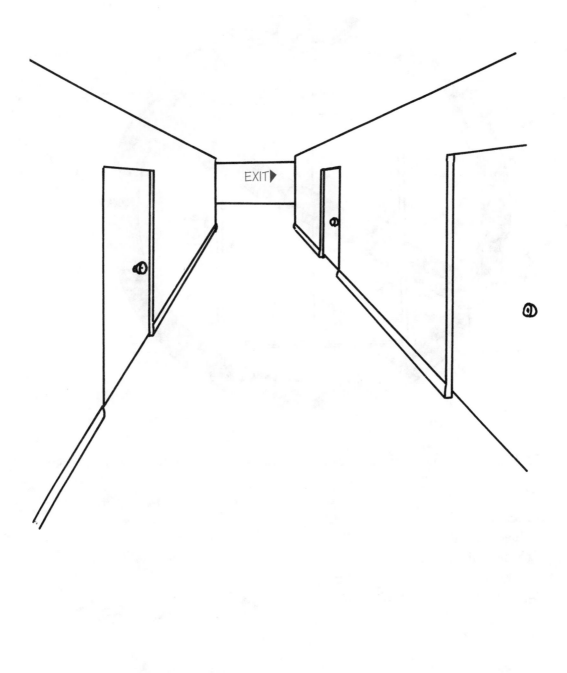

❏ If a fire breaks out, don't use the elevator. A power failure could trap you inside.

❏ When traveling, purchase a portable smoke detector to hang on your door.

Tools & materials

❏ Screwdriver
❏ Electric drill
❏ Small bit
❏ Pencil
❏ Hammer

Installing a smoke detector

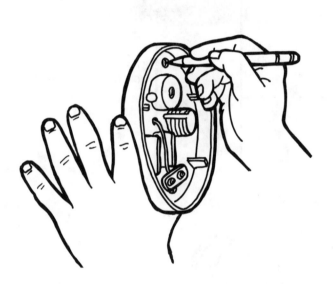

Step 6-1.
Marking the holes for the anchors.
Open the cover, place the base of the
detector against the wall or ceiling,
and mark the screw holes
with a pencil.

Step 6-2.
Drilling the pilot holes for the anchors.
Use an electric drill and a small bit to drill
pilot holes for the screw anchors.

Step 6-3.
Installing the
anchors and screws.
Use a hammer to gently tap the anchors into the holes so that they are flush with the surface. Screw the screws into the anchors, leaving the screw head sticking out about 1/8 of an inch.

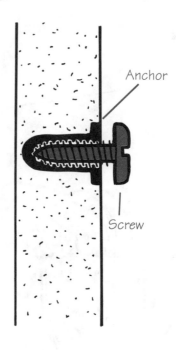

Anchor

Screw

Step 6-4.
Mounting the smoke dector.
Install a fresh battery and fit the base to the screws. Close the cover and test the alaram.

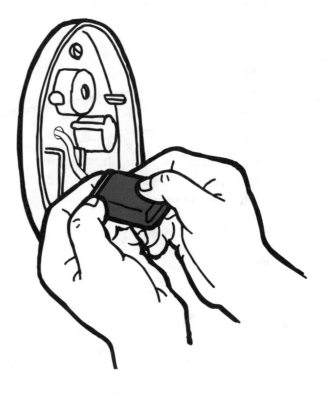

Installing a fire extinguisher

Step 6-5. Marking the holes for the bracket.

Mount the fire extinguisher near a doorway, about 5 feet above the floor. Try to find a stud or use screws and screw anchors to fasten the bracket to the wall. Hold the bracket against the wall at the desired location and use a pencil to mark the screw holes.

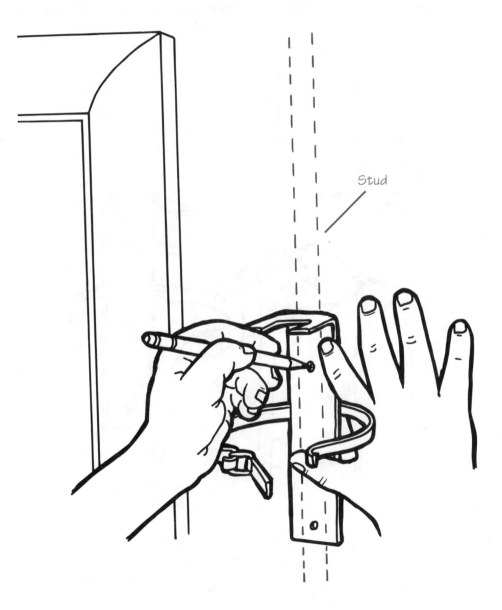

Step 6-6.
Mounting the bracket.
Use an electric drill and a small bit to drill pilot holes for the screws (for studs) or holes for the screw anchors. If you are using anchors, use a hammer to gently tap the anchors into the holes so that they are flush with the surface. Use the screws to fasten the bracket to the wall. Now fit the fire extinguisher into the bracket and close the retaining strap.

Remember, the fire extinguisher is useless unless you know how to use it. Memorize the word *PASS*—Pull, Aim, Squeeze, and Sweep. Pull the retaining pin, aim the extinguisher at the bottom of the fire, squeeze the handle to start the spray, and sweep the spray back and forth across the flame. Don't let the fire get between you and an exit. Stay low, and try not to breath any smoke or fumes. If after a couple of minutes you haven't contained or extinguished the flames, don't waste any more time. Close the windows and doors (if you can do so safely) to confine the fire, get out of the house, and call the fire department.

Escape procedures

Step 6-7.
Stay low
if there is smoke.
Crawl if necessary. If possible, cover your mouth and nose with a damp cloth. Move quickly to the exit door.

Step 6-8.
Check the door for heat.
Use the back of your hand. If the door is not hot, continue your exit quickly and cautiously.

Step 6-9. Don't open a hot door.

If the door is hot or if smoke is coming in from under the door, don't open the door. Go to a window. Open the window and use the chain ladder. If you can't open the window, break it out with a chair or other object, not your hands. If you have no ladder, carefully climb over the window sill and hang by your hands before dropping to the ground.

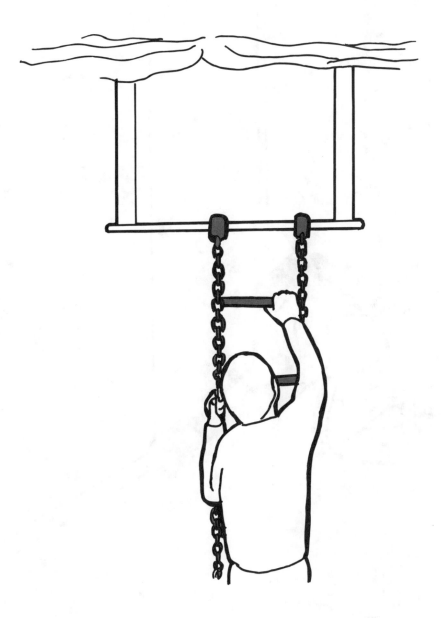

Step 6-10. Stop the flow of smoke.

If you're above the second story and the door is hot, stuff the crack around the door with sheets, towels, or curtains. Wet them if possible.

Step 6-11. Wet the walls.

Fill a waste basket with water and douse the door and walls with water. Stay in the room.

Step 6-12. Signal for help.

Go to the window and signal rescuers by waving a white cloth or flashlight. If you still have lights, turn the lights on and off rapidly.

CHAPTER SEVEN

First Aid

The goals of first aid are to help the injured person recover, or at least prevent the injury from becoming more serious, and to provide comfort and reassurance. First aid is usually all that is needed for most minor injuries; however, serious injuries should always be treated by qualified medical professionals.

The following steps have been recommended by medical professionals. This chapter provides you with some basic information, but you will be much better qualified to perform first aid if you attend classes that include instructions in cardiopulmonary resuscitation (CPR). CPR should be attempted only by trained personnel. The best advice in any emergency is to stay calm and call 911—they'll tell you what to do.

First aid kit

- ❏ Adhesive tape
- ❏ Sterile gauze bandage (2-x- 2-, 3-x-3-, and 4-x- 4-inch pads)
- ❏ Gauze roller bandage (roll about 15 feet long)
- ❏ Adhesive bandages (various sizes)
- ❏ Cotton swabs
- ❏ Tweezers
- ❏ Small scissors
- ❏ Calamine lotion
- ❏ Antibiotic ointment
- ❏ Antibiotic spray
- ❏ Rubbing alcohol (used to sterilize tweezers, etc.)
- ❏ Reusable ice pack or package of frozen peas or corn

Bites & stings

Minor insect bites or stings can cause pain, swelling, or redness at the bite, as well as itching or burning. Only the honey bee leaves the stinger and venom sac in the skin. The venom sac continues to pump venom for up to 3 minutes. Gently remove the stinger by scraping with your fingernail or a nail file, being careful not to squeeze the stinger or sac. Multiple stings can cause a toxic reaction, with symptoms such as headaches, cramps, fever, and even unconsciousness, and should be treated by trained professionals.

Most insect bites are normally only little welts that itch and go away in a couple of days; however, flies and mosquitoes can spread disease. Steps 7-1 through 7-3 apply to insect bites; Steps 7-4 through 7-10 cover what to do in the case of a cat or dog bite.

Step 7-1. Cleaning the area.
Wash the bite thoroughly
with soap and water, then
apply an antiseptic.

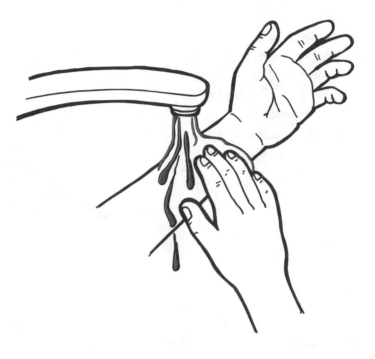

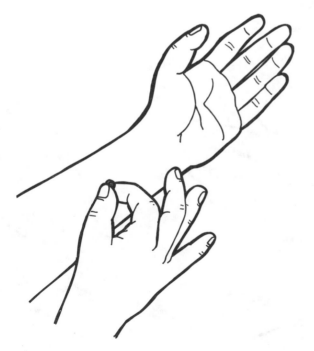

Step 7-2. Rubbing with aspirin.
As soon as possible, moisten the
skin and rub an aspirin tablet
gently over the bite.

Step 7-3.
Reducing the itch.
To reduce the itch of an insect bite, hold an ice cube or ice pack on the bite for about 5 minutes. Repeat three or four times, or try one or more of the following methods.

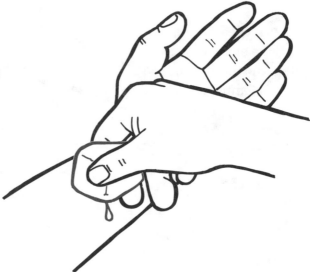

Method A.
Apply a paste of water and salt to the bite.

Method B.
Dissolve 1 tablespoon of baking soda in a glass of water. Dip a cloth in the solution and place on the bite for about 15 minutes.

Method C.
Apply calamine lotion to the bite.

Step 7-4.
Rinsing the bite.

Cat bites might not look serious, but they are more dangerous than dog bites. Cats' mouths carry much more bacteria. Hold the bite under cool or slightly warm running water for about 10 minutes.

Step 7-5. Washing the bite.

Carefully wash the bite with soap and water, then rinse under running water.

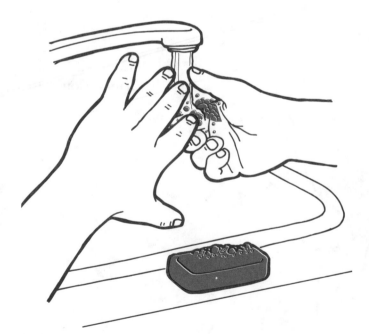

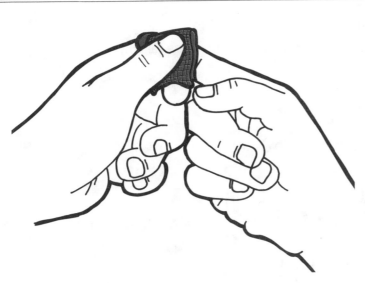

Step 7-6.
Removing debris.
Pat gently with a sterile gauze pad and remove any particles with sterile tweezers or a cotton swab. Rinse the bite again under running water for about 5 minutes. Then dry the area.

Step 7-7. Stopping the bleeding.
If the wound is still bleeding, place a sterile pad over the bite, apply direct pressure, and elevate the wound above heart level.

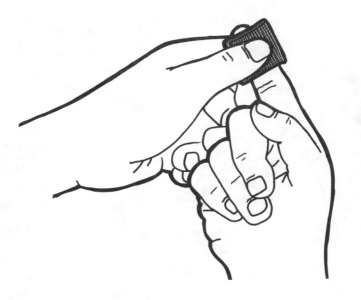

Step 7-8. Protecting the wound.

Apply a nonadherent dressing and cover with a sterile gauze fastened with adhesive tape.

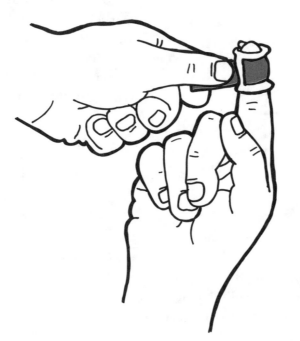

Step 7-9.
Checking for rabies.

Take the victim to the hospital immediately if you think the animal was rabid. The animal's behavior might not indicate any signs of rabies, so try to check the animal's collar for a rabies vaccination tag, if possible, or have the animal impounded for observation. If the animal cannot be found, go to the hospital so that precautionary measures can be taken.

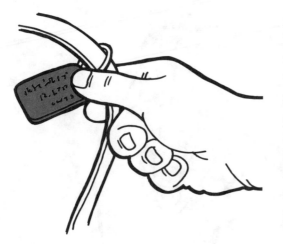

Step 7-10.
Getting medical help.
Seek professional help if the victim has not had a tetanus shot within the past 10 years, or if the bite becomes infected (gets red, painful, or hot). In the case of anything more than a minor bite, seek professional help.

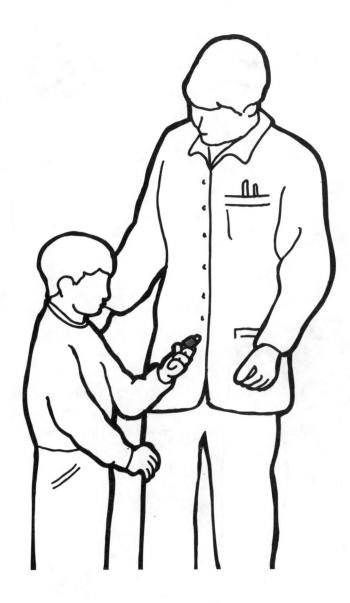

Bruises

Bruises are a common problem with active kids as well as adults. Steps 7-11 and 7-12 offer some tips.

Step 7-11. Applying ice.
Apply an ice pack as soon as possible after the injury. Keep the ice pack on the bruise for about 15 minutes at a time, then let the skin warm naturally.

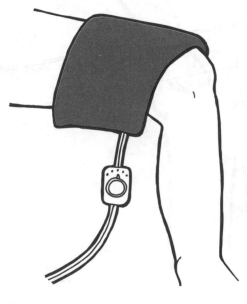

Step 7-12. Applying heat.
If you don't notice the bruise until a day or two after the injury, apply heat instead of ice.

Burns

The quick response required for a fire is the same needed for a burn—you must get the fire, or burn, out fast. Most first- and second-degree burns can be treated at home, but third-degree burns require professional medical attention.

First-degree burns include most sunburns and scalds that are usually red and painful. Second-degree burns include severe sunburns, burns caused by brief contact with hot cooking pots or heating elements, and other severe burns that blister, ooze, and are always painful. Third-degree burns are not usually painful because nerve endings have been destroyed. These burns are charred and light colored, and can be caused by longer contact with a hot surface, electricity, or chemicals. Third-degree burns always require treatment by a doctor.

Any burn that you have doubts about, one that has become infected, or one that hasn't healed in a couple of weeks should be treated by a doctor. If you are going to see a doctor about a burn, wash it, but don't use any ointments or antiseptics. Just cover it with a dry, sterile bandage.

Step 7-13. Stopping the burning process.

Flush the burned area with cold (not ice) water for about 30 minutes, or until the burning stops. If the burn was caused by hot grease, soup, or battery acid, first remove any saturated clothing, wash the grease or other contaminant off the skin, then soak the burn in cold water. If the clothing sticks to the burn, don't try to pull it off. Just rinse over the clothing and get professional help.

Step 7-14. Applying ointments.

DO NOT put any grease, ointment, or cream on the burn. Apply calamine lotion to large areas of mild sunburns. Just gently cover the burn in a dry, sterile cloth, and let it start to heal for a day or so.

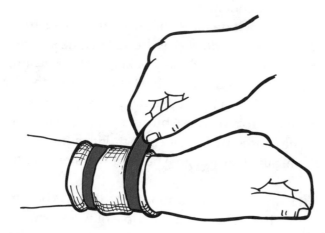

Step 7-15. Caring for blisters.

Leave blisters alone. Blisters are nature's best bandage. If a blister breaks, gently wash the area with soap and water, apply a dab of antibiotic ointment, and cover with a sterile bandage.

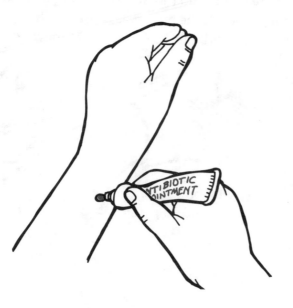

Step 7-16. Protecting the burn.

The next day, gently wash the burn with soap and water, dry the area, and keep it covered.

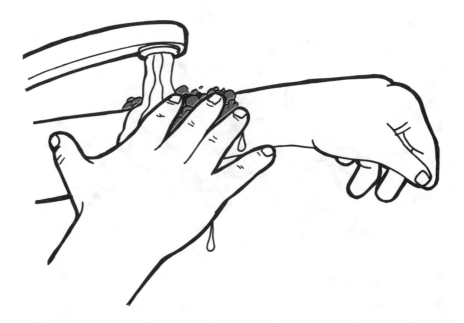

Step 7-17. Using vitamin E.

When the burn has started to heal, cut open a vitamin E capsule and rub the liquid onto the skin.

Step 7-18. Preventing infection.
To discourage infection as the burn heals, apply an over-the-counter antibiotic ointment.

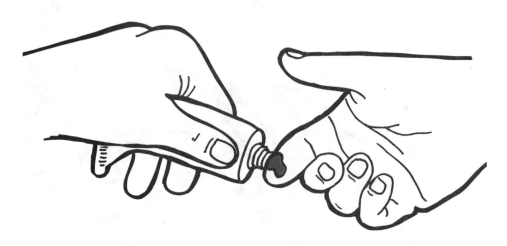

Cuts & scrapes

You slice your finger instead of the tomato, the wrench slips and you scrape your arm while doing a repair, or the tricycle overturns on the sidewalk and a small knee is skinned. Although painful, these minor cuts and scrapes are seldom dangerous and can be treated at home. Minor bleeding normally stops by itself within a few minutes; however, if blood spurts or flows from the wound so much that it can't be stopped after a few minutes of applied pressure, it is severe bleeding and you must get medical help immediately.

See a doctor if:

- The blood is bright red and spurting.
- The wound is large and deep.
- Foreign debris cannot be washed from the wound.
- Pus weeps, red streaks appear, or redness develops more than an inch around the wound.

For minor bleeding, follow Steps 7-19 and 7-20. For severe bleeding, follow Steps 7-21 and 7-22.

Step 7-19. Cleaning the wound.

Gently wash the area with soap and water or just water. Hold it under running water for a few minutes to flush out any dirt or grit. Remove any stubborn debris with a cotton swab or sterile tweezers.

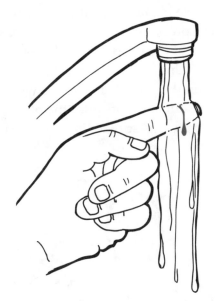

Step 7-20.
Dressing the wound.

Apply an antibiotic cream or spray, then cover the scrape with an adhesive bandage or sterile gauze pad and adhesive tape. For a small cut, use a butterfly bandage or regular adhesive bandage to close the cut. Gently apply the bandage across the cut to close, but not overlap, the edges.

Step 7-21. Stopping the blood flow.

Place a clean, absorbent cloth, a sterile bandage, or even a towel over the wound and press firmly. If a cloth isn't handy, just use your hand. The bleeding should stop in a minute or two. If blood continues to seep through the first cloth, add a second and third cloth as necessary. Don't remove the first one because it could dislodge some of the clotting blood. Continue applying pressure. If the bleeding doesn't stop, raise the wound above the level of the heart. If it still doesn't stop, get medical help immediately. Until help arrives, use a pressure point to slow the bleeding. Press on the point between the wound and the heart. Hold pressure for about a minute after the bleeding stops. If bleeding starts again, reapply pressure to the pressure point. DO NOT use a tourniquet.

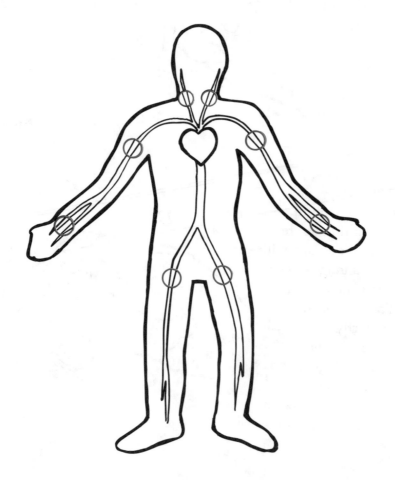

Step 7-22.
Dressing the wound.
When the bleeding has stopped, apply an over-the-counter antibiotic ointment and cover with a sterile gauze bandage. Make sure the victim's tetanus shot is up to date.

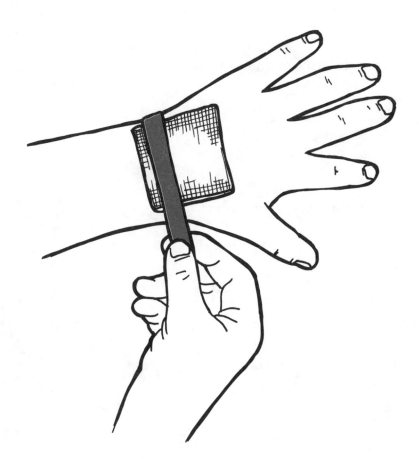

Glossary

case A protective cover or housing for a lock.

cylinder The turning part of a lock.

deadbolt lock
A lock in which the bolt
is operated by hand or a key.

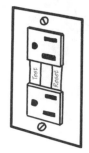

ground-fault circuit interrupter (GFCI)
A safety device installed in a circuit designed
to prevent electric shock by detecting very
small currents and interrupting the flow. GFCI
receptacles are used in damp areas, such as
bathrooms and basements.

hole saw
A cutting cylinder attached to a drill bit, designed to cut large-diameter holes.

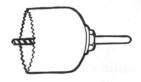

hollow-core door A door made of two panels with an empty space between the panels.

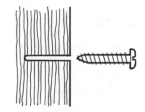

jamb The side part of a door frame.

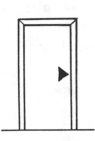

pilot hole A guide hole for screws, usually formed by tapping in a nail.

rim lock A lock with a vertical deadbolt designed to be mounted on the surface of the door.

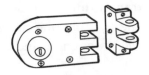

solid-core door   A door constructed so that its interior is completely filled.

stake out   The stalking or surveillance of a property or person.

strike plate   The metal plate fastened to the door jamb that holds the bolt and secures the door.

template   A pattern used for forming an accurate copy.

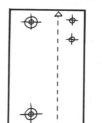

transformer   A device used to transfer electrical energy from one circuit to another using electromagnetic induction.

voltage tester   A device, or meter, used to test for or measure voltage.

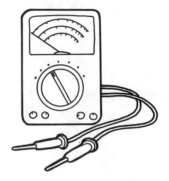

wire nut   A solderless connector used to connect two or more wires together to form an electrical connection.

wood dowel   A wooden plug or pin normally fitted in holes to fasten two pieces together.

Index

A

ABC-class fire extinguisher, 92
amateur burglars, 2
antilift screws or dowels, 52-53, 65-67
antislide locks, 63
automatic outdoor light, 19-22
  mounting, 22
  old light removal, 19-20
  old wires, disconnecting, 20
  voltage check, 20
  wire connection, 21
  wiring, completion of, 21

B

bolt lock, 68-71
broom handle, 63
burglar alarms, 72-89
  basic parts, 73
  door or window alarm, 78-88
  doorknob, 77
  magnets, 74
  motion-sensing alarm, 75-76
  self-contained alarm, 77
  tools and materials, 76
  types, 72
burglars, 1-12
  amateur, 2
  neighborhood kid, 3
  neighborhood watch program, 8
  professional, 3-4
  profiles of, 1-12
  safety from, 10-12
  security check, 4-10
  semiprofessional, 2-3

C

chisel, using, 41

D

deadbolt lock, 38-45, 57-59
  adjusting lock to door, 42
  bolt, marking for, 44
  chisel, using, 41
  drilling hole for bolt, 40
  drilling hole for lock, 39
  face plate, marking, 40
  installing, 38-45
  lock assembly, installing, 42
  marking holes, 39
  retaining plate, installing, 43
  screws, installing, 41
  sliding glass doors, 57-59
  strike plate,
    installing, 45
    drilling for, 45
    marking for, 44
  template, 38
  thumbturn, installing, 43
door or window alarm, 78-89
  control box, 80-82
    anchors and screws, installation, 81
    anchors, drilling pilot holes, 81
    template, marking, 80
    leveling, 80
    mounting, 82
  drill bit, starting points, 78
  magnet, installing, 79
  pilot holes, drilling, 78
  sensor, connecting, 84
  sensor and magnet, marking holes, 87
  sensor and magnet, mounting, 88

template, using, 78
transformer, connecting, 83
transmitter, 85-86
  mounting, 85-86
  programming, 84
  wireless alarm, installing, 79
  wires, stripping, 82
door pinning, 36-37
  installation, 36
doorknob alarm, 77
doors and locks, 32-61
  deadbolt lock, 38-45
  door pinning, 36-37
  rim lock, 46-50
  screw holes, repairing, 60-61
  sliding glass doors, 51-61, 51
  tools and materials, 34
  wide-angle viewer, 33-35
doors, solid core, 32
double-hung windows, 68-71
  bolt lock,
    bolt hole, marking, 70
    completing job, 71
    holes, drilling, 71
    pilot holes, drilling, 69
    screw holes, marking, 68
    strike plate, marking, 70
    installing, 68-71
  pins, installing, 68

E

engraving valuables, 9

F

fire extinguisher, 90, 92, 99-101
  ABC-class, 92
  installing, 99-100
  use, 101
fire safety checklist, 93-96
  chain ladders, 94
  designated place of

assembly, 93
  fire exits, apartments, 95
fire escape procedures, 102-106
first aid, 107-123
  bites and stings, 107-115
  bruises, 116
  burns, 117-120
  cuts and scrapes, 120-123
  first aid kit list, 108

G

ground-fault circuit interrupter (GFCI), 28-31
  hot wire, finding, 29
  installation, 28-31
  mounting, 31
  old receptacle, removal of, 28
  receptacle, examining, 29
  wires, connecting, 30
    one set, 30
    two sets, 30

L

landscaping, 14-15
low-voltage yard light, 23-27
  cable connection, 25
  cable, burying, 27
  installing, 23-27
  laying cable, 24
  light connection, 26
  placement, 27
  positioning, 24
  transformer installation, 23

M

motion-sensing alarm, 75-76

N

neighborhood kid burglars, 3
neighborhood watch program, 8

O

outdoor lighting, 12-31
  automatic outdoor light, 19-22
  connecting wires,
    screw terminal, 18
    wire nuts, 17
  ground-fault circuit interrupter (GFCI), 28-31
  low-voltage, 14, 23-27
    yard light, 23-27
  tools and materials, 16
  using wire stripper, 16
  voltmeter, using, 18

P

photographing valuables, 9
professional burglars, 3-4

R

rim lock, 46-50
  completion, 50
  door trim, cutting, 49
  drilling holes, 47
  installing, 46-50
  jamb,
    cutting, 50
    marking, 49
  lock installation, 47
  strike plate, marking, 48
  template, marking holes with, 46

S

safety, 10-12
  bathroom, 11
  bedrooms, 11
  children's bedrooms, 11
  garage or basement, 11-12
  hallways and stairs, 10-11
  kitchen, 10
  living room, 10
security bar, 53-56
  adjustment, 55
  completion, 56
  hinge bracket, installing, 54

security bar (*cont.*)
  installing, 53-56
  leveling, 55
  pilot holes, drilling, 54
  saddle bracket, marking
    holes, 56
security check, 4-10
self-contained alarm, 77
semiprofessional burglars,
  2-3
sliding glass doors, 51-61
  antilift screws, 52-53
  deadbolt lock, 52, 56-59
    bolt hole, marking, 58
    completion, 59
    drill starting point,
    marking, 59
    pilot holes, 57
    ventilation, marking for,
    58
  security bar, 53-56

sliding windows, 65-67
  antilift screws or dowels,
    65-67
    dowels, installing, 66
    pilot holes, drilling, 65
    screws, installing, 66
  pins, installing, 67
  securing, 65-67, 65
smoke detectors, 89-90
  installing, 96-97
  placement, 90
  tools and materials, 95

T

transformer installation, 23

V

valuables
  engraving, 9
  photographing, 9

volt meter, 18

W

wide-angle viewer, 33-35
  installing, 34-35
windows, 62-71
  antilift screw or dowel, 64
  antislide locks, 63
  broom handle, 63
  double-hung, 68-71, 68
  sliding window, 65-67,
    65
  tools and materials, 65
wire nuts, 17, 26
wire stripper, using, 16

Y

yard light, low-voltage, 22-
  27
yard lighting, 6